SpringerBriefs in Molecular Science

Green Chemistry for Sustainability

Series editor

Sanjay K. Sharma, Jaipur, India

More information about this series at http://www.springer.com/series/10045

Pratima Bajpai

Pretreatment of Lignocellulosic Biomass for Biofuel Production

Pratima Bajpai
Pulp and Paper Consultants
Patiala, Punjab
India

ISSN 2191-5407 ISSN 2191-5415 (electronic)
SpringerBriefs in Molecular Science
ISSN 2212-9898
SpringerBriefs in Green Chemistry for Sustainability
ISBN 978-981-10-0686-9 ISBN 978-981-10-0687-6 (eBook)
DOI 10.1007/978-981-10-0687-6

Library of Congress Control Number: 2016933107

Printed on acid-free paper

This Springer imprint is published by Springer Nature
The registered company is Springer Science+Business Media Singapore Pte Ltd.

Preface

Rising oil prices and uncertainty over the security of existing fossil fuel reserves, combined with concerns over global climate change, have created the need for new transportation fuels and bioproducts to substitute for fossil carbon-based materials. Ethanol is considered to be the next-generation transportation fuel with the most potential, and significant quantities of ethanol are currently being produced from corn and sugarcane via a fermentation process. Utilizing lignocellulosic biomass as a feedstock is seen as the next step toward significantly expanding ethanol production. The biological conversion of cellulosic biomass into bioethanol is based on the breakdown of biomass into aqueous sugars using chemical and biological means, including the use of hydrolytic enzymes. From that point, the fermentable sugars can be further processed into ethanol or other advanced biofuels. Therefore, pretreatment is required to increase the surface accessibility of carbohydrate polymers to hydrolytic enzymes. The goal of the pretreatment process is to break down the lignin structure and disrupt the crystalline structure of cellulose, so that the acids or enzymes can easily access and hydrolyze the cellulose. Pretreatment can be the most expensive process in biomass-to-fuels conversion but it has great potential for improvements in efficiency and lowering of costs through further research and development. Pretreatment is an important tool for biomass-to-biofuels conversion processes and is the subject of this e-book.

Contents

List of Figures

List of Tables

Chapter 1
Background and General Introduction

Abstract Owing to the increasing demand for energy, the hunt for alternative sources of energy generation that are inexpensive, ecofriendly, renewable, and can replace fossil fuels is on. One approach in this direction is the conversion of plant residues into biofuels wherein lignocellulose, which forms the structural framework of plants consisting of cellulose, hemicellulose, and lignin, is first broken down and hydrolyzed into simple fermentable sugars, which upon fermentation form biofuels. A major bottleneck is to disarray lignin which is present as a protective covering and makes cellulose and hemicellulose recalcitrant to enzymatic hydrolysis. A number of biomass pretreatment processes have been used to break the structural framework of plants and to depolymerize lignin. In this chapter, background and general introduction on pretreatment of lignocellulosic biomass for biofuel production are presented.

Keywords Energy · Biofuels · Lignocellulose · Cellulose · Hemicelluloses · Lignin · Fermentation · Biomass · Pretreatment · Enzymatic hydrolysis

Development of sustainable energy systems based on renewable biomass feedstocks is now a universal effort. Biofuels produced from various lignocellulosic materials, such as wood, agricultural, or forest residues, have the potential to be a valuable substitute for, or complement to, gasoline. Liquid biofuels, such as ethanol, can be produced from biomass via fermentation of sugars produced from the cellulose and hemicellulose within the lignocellulosic materials. Ethanol produces slightly less greenhouse emissions than fossil fuel (carbon dioxide is recycled from the atmosphere to produce biomass); can replace harmful fuel additives (e.g., methyl tertiary butyl ether), and produces jobs for farmers and refinery workers (Bajpai 2013). It is easily applicable in present-day internal combustion engine vehicles (ICEVs), as mixing with gasoline is possible. Ethanol is already commonly used in a 10 % ethanol/90 % gasoline blend. Adapted ICEVs can use a blend of 85 % ethanol/15 % gasoline (E85) or even 95 % ethanol (E95). Ethanol addition increases octane and reduces carbon monooxide, volatile organic carbon and particulate emissions of gasoline. And, via on-board reforming to hydrogen, ethanol is

P. Bajpai, *Pretreatment of Lignocellulosic Biomass for Biofuel Production*,
SpringerBriefs in Green Chemistry for Sustainability,
DOI 10.1007/978-981-10-0687-6_1

also suitable for use in future fuel cell vehicles (FCVs). These vehicles are believed to have about double the current ICEV fuel efficiency.

Use of lignocellulosic biomass as a feedstock is seen as the next step toward significantly expanding ethanol production. Lignocellulosic biomass includes the following:

- Forestry wastes (e.g., wood chips, and sawdust)
- Agricultural residues (e.g., corn stover (cob and stalk), rice straw, bagasse, cotton gin trash, etc.)
- Bioenergy crops (sweet sorghum, switchgrass and common reeds)
- Industrial wastes (e.g., paper sludge, recycled newspaper)
- Municipal solid wastes.

Unlike food-based (starch-derived) biomass, it shows a series of advantages such as, low cost, abundant supplies, non-competition with grain as food (Sathisuksanoh et al. 2009).

Lignocellulose, a natural complicated composite basically consists of cellulose, hemicellulose, and lignin. Cellulose and hemicellulose are tangled together and wrapped by lignin outside (de Vries and Visser 2001). Depending on sources and cell types, the dry weight typically makes up of around 35–50 % cellulose, 20–35 % hemicellulose, and 10–25 % lignin (Demirbas 2005). Cellulose, the most abundant natural carbon bioresource on the earth, is a homopolysaccharide of anhydroglucopyranose linked by ß-1, 4-glycosidic linkages (McMillan 1997). Adjacent cellulose chains are coupled via orderly hydrogen bonds and Van der Waal's forces, resulting in a parallel alignment and a crystalline structure (Zhang et al. 2007). Several elementary fibrils gather, forming much larger microfibrils, which are further bundled into larger macrofibrils, leading to the rigidity and strength of cell walls. Efficient conversion of cellulose into glucose has been a central topic for long. Hemicellulose, the second main polysaccharide, is a polymer containing primarily pentoses (xylose and arabinose) with hexoses (glucose and mannose), which are dispersed throughout and form a short-chain polymer that intertwines with cellulose and lignin like a glue (Wilkie 1979). Lignin is a polymer consisting of various aromatic groups. It can be converted into numerous chemical products that are made from fossil-based chemical industry, including coal, oils and natural gas. The production of cellulosic ethanol holds promise for an improved strategic national security, job creation, strengthened rural economies, improved environmental quality, nearly zero net greenhouse gas emissions, and sustainable local resource supplies (Demain et al. 2005; Lynd et al. 1991, 1999, 2002; Zhang 2008). So far, more companies are working on reducing costs of cellulosic ethanol production, such as Iogen Corporation, Abengoa Bioenergy, Dupont, British Petroleum, Mascoma (Biotechnology Industry Organization 2008). The price of E85 is still as high as $1.81 per gallon (DOE 2009).

Lignocellulosic biomass can be converted into soluble sugars via a large number of approaches. Conversion for lignocellulose is much more complicated and difficult, considering their complex and recalcitrant structures (Jorgensen et al. 2007; Yang and Wyman 2008). The biological conversion of cellulosic biomass into bioethanol is based on the breakdown of biomass into aqueous sugars using chemical and biological means, including the use of hydrolytic enzymes. From that point, the fermentable sugars can be further processed into ethanol or other advanced biofuels. Therefore, pretreatment is required to increase the surface accessibility of carbohydrate polymers to hydrolytic enzymes. The goal of the pretreatment process is to break down the lignin structure and disrupt the crystalline structure of cellulose, so that the acids or enzymes can easily access and hydrolyze the cellulose. Pretreatment can be the most expensive process in biomass-to-fuels conversion but it has great potential for improvements in efficiency and lowering of costs through further research and development. Pretreatment is an important tool for biomass-to-biofuels conversion processes. Numerous lignocellulose pretreatment approaches have been developed, which are reviewed by many researchers (Chandra et al. 2007; Dale 1985; Eggeman and Elander 2005; Himmel et al. 2007; Ladisch et al. 1992; Lin et al. 1981; Lynd 1996; Lynd et al. 2003, 2008; McMillan 1994; Mosier et al. 2005; Office of Energy Efficiency and Renewable Energy and Office of Science 2006; Ragauskas et al. 2006; Sun and Cheng 2002; Vertès et al. 2006; Wyman 2007; Wyman et al. 2005a, b; Bensah and Mensah 2009; Zheng et al. 2009; Chaturvedi and Verma 2013; Agbor et al. 2011; Kumar et al. 2009; Avgerinos and Wang 1983). Most notably, a collaborative team called consortium for applied fundamentals and innovation (CAFI) funded by the Department of Energy and Department of Agriculture has formed and focused on several "leading pretreatment technologies," including dilute (sulfuric) acid pretreatment, flow-through pretreatment, ammonia fiber expansion (AFEX), ammonia recycle percolation (ARP), and lime pretreatment for the past several years (Moxley 2007).

References

Agbor VB, Cicek N, Sparling R, Berlin A, Levin DB (2011) Biomass pretreatment: fundamentals toward application. Biotechnol Adv 29:675–685

Avgerinos GC, Wang DIC (1983) Selective delignification for fermentation of enhancement. Biotechnol Bioeng 25:67–83

Bajpai P (2013) Advances in bioethanol. SpringerBriefs in Applied Science and Technology Springer Science and business media, New York

Bensah EC, Mensah M (2009) Chemical pretreatment methods for the production of cellulosic ethanol: technologies and innovations Hindawi Publishing Corporation. Int J Chem Eng 2013, 21 p, Article ID 719607. http://dx.doi.org/10.1155/2013/719607

Biotechnology Industry Organization (2008) Achieving sustainable production of agricultural biomass for biorefinery feedstock. http://www.bio.org/ind/biofuel/sustainablebiomassreport.pdf

Chandra R, Bura R, Mabee W, Berlin A, Pan X, Saddler J (2007) Substrate pretreatment: the key to effective enzymatic hydrolysis of lignocellulosics? Adv Biochem Eng Biotechnol 108:67–93

Chaturvedi V, Verma P (2013) An overview of key pretreatment processes employed for bioconversion of lignocellulosic biomass into biofuels and value added products. 3 Biotech 3:415–431. doi:10.1007/s13205-013-0167-8

Dale BE (1985) Cellulose pretreatment: technology and techniques. Annu Rev Ferment Proc 8:299–323 de Vries RP, Visser J (2001) Aspergillus enzymes involved in degradation of plant cell wall polysaccharides. Microbiol Mol Biol Rev 65:497–522

Demain AL, Newcomb M, Wu JHD (2005) Cellulase, clostridia, and ethanol. Microbiol Mol Biol Rev 69:124–154

Demirbas A (2005) Bioethanol from cellulosic materials: a renewable motor fuel from biomass. Energy Sour 27(4):327–337

DOE BP (2009) Biomass program. http://www1.eere.energy.gov/biomass/

Eggeman T, Elander RT (2005) Process and economic analysis of pretreatment technologies. Biores Technol 96:2019–2025

Himmel ME, Ding S-Y, Johnson DK, Adney WS, Nimlos MR, Brady JW, Foust TD (2007) Biomass recalcitrance: engineering plants and enzymes for biofuels production. Science 315 (5813):804–807

Jorgensen H, Kristensen JB, Felby C (2007) Enzymatic conversion of lignocellulose into fermentable sugars: challenges and opportunities. Biofuels Bioprod Biorefin 1(2):119–134

Kumar P, Barrett DM, Delwiche MJ, Stroeve P (2009) Methods for pretreatment of lignocellulosic biomass for efficient hydrolysis and biofuel production. Ind Eng Chem Res 48:3713–3729

Ladisch MR, Waugh RL, Westgate P, Kohlmann K, Hendrickson R, Yang Y, Ladisch C (1992) Intercalation in the pretreatment of cellulose. ACS Symp Ser 515:509–519

Lin KW, Ladisch MR, Schaeffer DM, Noller CH, Lechtenberg V, Tsao GT (1981) Review on effect of pretreatment on digestibility of cellulosic materials. AIChE Sym Ser 203:102–106

Lynd LR (1996) Overview and evaluation of fuel ethanol from cellulosic biomass: technology, economics, the environment, and policy. Annu Rev Energy Env 21:403–465

Lynd LR, Cushman JH, Nichols RJ, Wyman CE (1991) Fuel ethanol from cellulosic biomass. Science 251:1318–1323

Lynd LR, Wyman CE, Gerngross TU (1999) Biocommodity engineering. Biotechnol Prog 15:777–793

Lynd LR, Weimer PJ, van Zyl WH, Pretorius IS (2002) Microbial cellulose utilization: fundamentals and biotechnology. Microbiol Mol Biol Rev 66:506–577

Lynd LR, Jin H, Michels JG, Wyman CE, Dale B (2003) Bioenergy: background, potential, and policy. Center for Strategic and International Studies, Washington, D.C.

Lynd LR, Laser MS, Bransby D, Dale BE, Davison B, Hamilton R, Himmel M, Keller M, McMillan JD, Sheehan J et al (2008) How biotech can transform biofuels. Nat Biotechnol 26 (2):169–172

McMillan JD (1994) Pretreatment of lignocellulosic biomass. In: Himmel ME, Baker JO, Overend RP (eds) Enzymatic conversion of Biomass for Fuels production. American Chemical Society, Washington, pp 292–324

McMillan JD (1997) Bioethanol production: status and prospects. Renew Energy 10(2–3):295–302

Mosier N, Wyman CE, Dale BE, Elander RT, Lee YY, Holtzapple M, Ladisch M (2005) Features of promising technologies for pretreatment of lignocellulosic biomass. Biores Technol 96:673–686

Moxley GM (2007) Studies of cellulosic ethanol production from lignocellulose. Virginia Tech, Blacksburg, p 78

Office of Energy Efficiency and Renewable Energy, Office of Science (2006) Breaking the biological barriers to cellulosic ethanol: a joint research agenda. A research roadmap resulting from the biomass to biofuels workshop. http://www.doegenomestolife.org/biofuels/

Ragauskas AJ, Williams CK, Davison BH, Britovsek G, Cairney J, Eckert CA, Frederick WJ Jr, Hallett JP, Leak DJ, Liotta CL (2006) The path forward for biofuels and biomaterials. Science 311(5760):484–489

Sathisuksanoh N, Zhu Z, Rollin J, Zhang Y-H (2009) Cellulose solvent- and organic solvent-based lignocellulose fractionation (COSLIF). In: Waldron K (ed) Bioalcohol production. Woodheading publishing, UK

Sun Y, Cheng J (2002) Hydrolysis of lignocellulosic materials for ethanol production: a review. Biores Technol 83:1–11

Vertès AA, Inui M, Yukawa H (2006) Implementing biofuels on a global scale. Nat Biotechnol 24:761–764

Wilkie K (1979) The hemicelluloses of grasses and cereals. Adv Carbohydr Chem Biochem 36 (1):215–264

Wyman CE (2007) What is (and is not) vital to advancing cellulosic ethanol. Trends Biotechnol 25 (4):153–157

Wyman CE, Dale BE, Elander RT, Holtzapple M, Ladisch MR, Lee YY (2005a) Comparative sugar recovery data from laboratory scale application of leading pretreatment technologies to corn stover. Biores Technol 96:2026–2032

Wyman CE, Dale BE, Elander RT, Holtzapple M, Ladisch MR, Lee YY (2005b) Coordinated development of leading biomass pretreatment technologies. Biores Technol 96:1959–1966

Yang B, Wyman CE (2008) Pretreatment: the key to unlocking low-cost cellulosic ethanol. Biofuels Bioprod Biorefin 2(1):26–40

Zhang Y-HP (2008) Reviving the carbohydrate economy via multi-product biorefineries. J Ind Microbiol Biotechnol 35(5):367–375

Zhang Y-HP, Ding S-Y, Mielenz JR, Elander R, Laser M, Himmel M, McMillan JD, Lynd LR (2007) Fractionating recalcitrant lignocellulose at modest reaction conditions. Biotechnol Bioeng 97(2):214–223

Zheng Y, Pan Z, Zhang R (2009) Overview of biomass pretreatment for cellulosic ethanol production. Int J Agric Biol Eng 2:51–68

Chapter 2
Structure of Lignocellulosic Biomass

Abstract Lignocellulosic materials consist mainly of three polymers: cellulose, hemicellulose, and lignin. These polymers are associated with each other in a hetero-matrix to different degrees and varying relative composition depending on the type, species, and even source of the biomass. The relative abundance of cellulose, hemicellulose, and lignin are inter alia, key factors in determining the optimum energy. Structural and compositional features of lignocellulosic biomass are presented in this chapter.

Keywords Lignocellulosic material · Polymers · Cellulose · Hemicelluloses · Lignin · Biomass · Structure · Composition

Plant biomass consists mainly of three polymers: cellulose, hemicellulose and lignin along with smaller amounts of pectin, protein, extractives and ash. The composition of these constituents can vary from one plant species to another. Hardwoods for example have greater amounts of cellulose, whereas wheat straw and leaves have more hemicellulose (Table 2.1). The ratios between various constituents within a single plant vary with age, stage of growth, and other conditions. These polymers are associated with each other in a heteromatrix to different degrees and varying relative composition depending on the type, species, and even source of the biomass (Carere et al. 2008; Chandra et al. 2007; Fengel and Wegener 1984). The relative abundance of cellulose, hemicellulose, and lignin are inter alia, key factors in determining the optimum energy conversion route for each type of lignocellulosic biomass (Mckendry 2002). Lignocellulosic feedstocks need aggressive pretreatment to yield a substrate easily hydrolyzed by commercial cellulolytic enzymes, or by enzyme producing microorganisms, to liberate sugars for fermentation. Cellulose is the main constituent of plant cell wall conferring structural support and is also present in bacteria, fungi, and algae. When existing as unbranched, homopolymer, cellulose is a polymer of beta-D-glucopyranose moieties linked via beta-(1,4) glycosidic bonds with well documented polymorphs (Fig. 2.1). The degree of polymerization of cellulose chains in nature ranges from 10,000 glucopyranose units in wood to 15,000 in native cotton. The repeating unit of the cellulose chain is the disaccharide

P. Bajpai, *Pretreatment of Lignocellulosic Biomass for Biofuel Production*,
SpringerBriefs in Green Chemistry for Sustainability,
DOI 10.1007/978-981-10-0687-6_2

Table 2.1 Cellulose, hemicellulose, and lignin contents in lignocellulosic biomass

	Cellulose	Hemicellulose	Biomass
Hardwoods	40–55	24–40	18–25
Softwoods	45–50	25–35	25–35
Wheat straw	30	50	15
Corn cobs	45	35	15
Grasses	25–40	35–50	10–30
Switchgrass	45	31.4	12

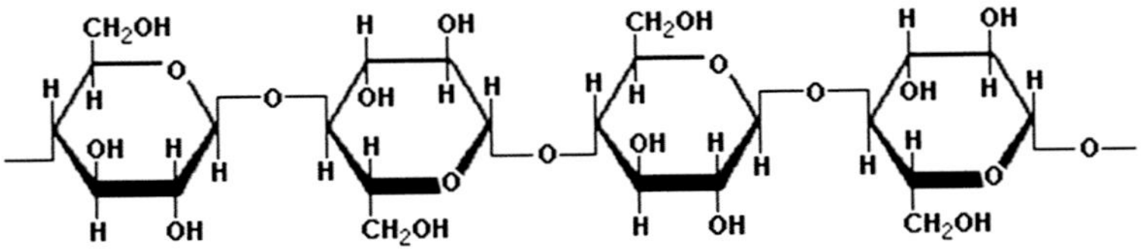

Fig. 2.1 Structure of cellulose

cellobiose as oppose to glucose in other glucan polymers (Desvaux 2005; Fengel and Wegener 1984). The cellulose chains (20–300) are grouped together to form microfibrils, which are bundled together to form cellulose fibers. The long-chain cellulose polymers are linked together by hydrogen and van der Waals bonds, which cause the cellulose to be packed into microfibrils. Hemicelluloses and lignin cover the microfibrils. Fermentable D-glucose can be produced from cellulose through the action of either acid or enzymes breaking the beta-(1,4)-glycosidic linkages. In biomass cellulose is present in both crystalline and amorphous forms. Crystalline cellulose contains the major proportion of cellulose, whereas a small percentage of unorganized cellulose chains form amorphous cellulose. Cellulose in its amorphous form is more susceptible to enzymatic degradation. The cellulose microfibrils are mostly independent but the ultrastructure of cellulose is largely due to the presence of covalent bonds, hydrogen bonding and Van der Waals forces. Hydrogen bonding within a cellulose microfibril determines 'straightness' of the chain but interchain hydrogen bonds might introduce order (crystalline) or disorder (amorphous) into the structure of the cellulose (Laureano-Perez et al. 2005).

Hemicellulose is the second most abundant polymer containing about 20–50 % of lignocellulose biomass. It is not chemically homogeneous like cellulose (Fig. 2.2). Hemicellulose has branches with short lateral chains consisting of different types of sugars. These monosaccharides include pentoses (xylose, rhamnose, and arabinose), hexoses (glucose, mannose, and galactose), and uronic acids (4-*O*-methylglucuronic, D-glucuronic, and D-galactouronic acids). The backbone of hemicellulose is either a homopolymer or a heteropolymer with short branches linked by beta-(1,4)-glycosidic bonds and occasionally beta-(1,3)-glycosidic bonds. Also, hemicelluloses can have some degree of acetylation, for example, in heteroxylan. Hemicelluloses have lower molecular weight compared to cellulose and branches with short lateral chains that are easily hydrolysed (Fengel and

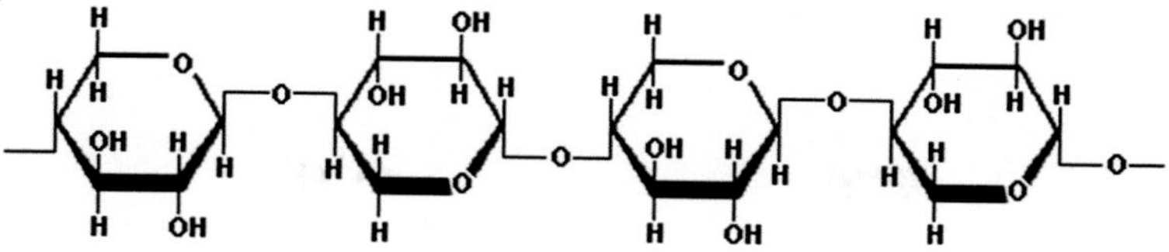

Xylan

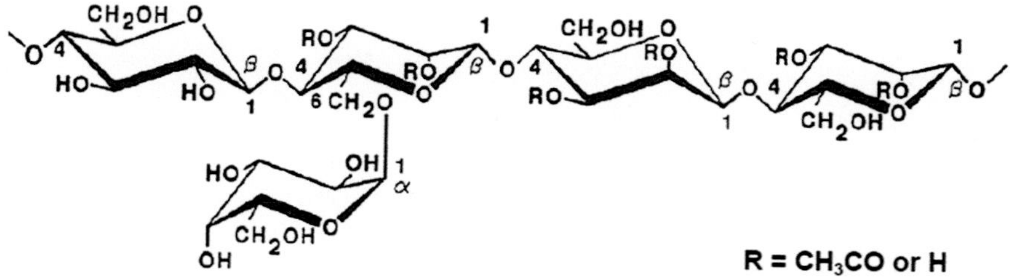

$R = CH_3CO$ or H

Galactoglucomannans

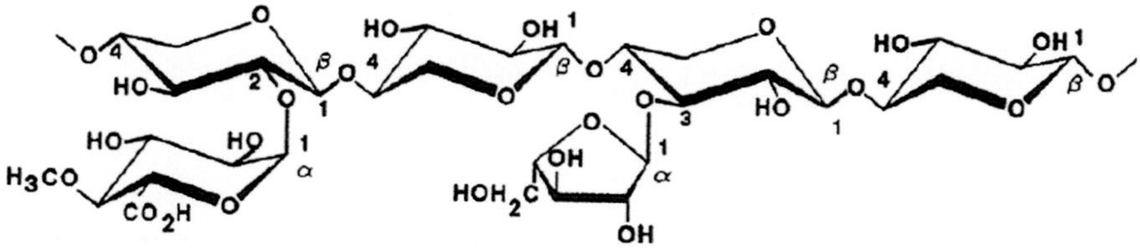

Arabinoglucuronoxylan

Fig. 2.2 Structure of hemicelluloses

Wegener 1984; Saha 2003). Hemicelluloses are found to differ in composition. In agricultural biomass like straw and grasses, hemicelluloses are composed mainly of xylan while softwood hemicelluloses contain mainly glucomannan (Fig. 2.2). In many plants, xylans are heteropolysaccharides with backbone chains of 1,4- linked beta-D-xylopyranose units. In addition to xylose, xylan may contain arabinose, glucuronic acid, or its 4-*O*-methyl ether, acetic acid, ferulic and p-coumaric acids. Xylan can be extracted easily in an acid or alkaline environment while extraction of glucomannan requires stronger alkaline environment (Balaban and Ucar 1999; Fengel and Wegener 1984). Among the key components of lignocellulosics, hemicelluloses are the most thermo-chemically sensitive (Hendricks and Zeeman 2009; Levan et al. 1990). Hemicelluloses within plant cell walls are thought to 'coat' cellulose-fibrils and it has been proposed that at least 50 % of hemicellulose should be removed to significantly increase cellulose digestibility. Nevertheless, severity parameters should be carefully optimized to avoid the formation of hemicellulose degradation products such as furfurals and hydroxymethyl furfurals

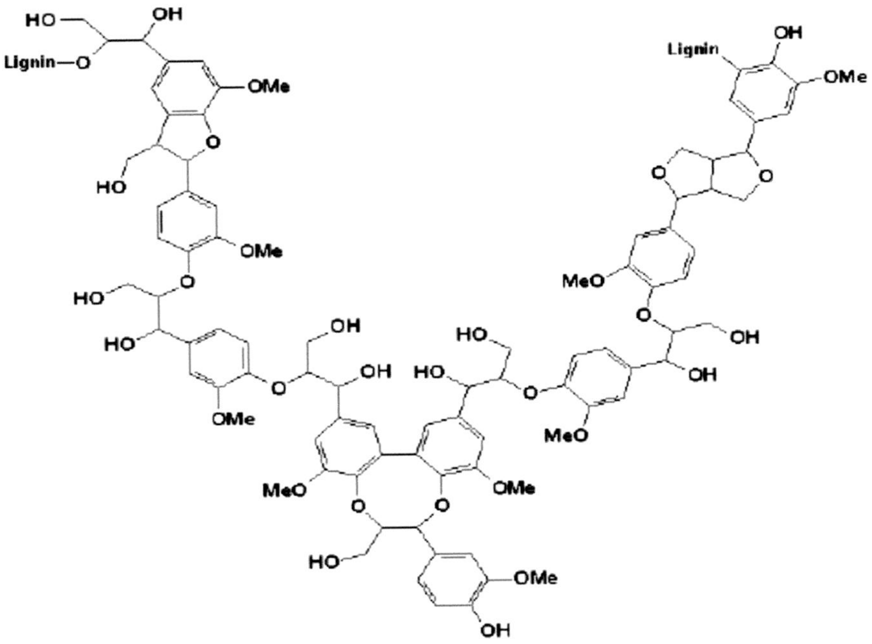

Fig. 2.3 Structure of lignin (complex cross-linked polymer of aromatic rings). Based on Walker (2010)

which have been reported to inhibit the fermentation process (Palmqvist and Hahn-Hägerdal 2000a, b). For this reason, pretreatment severity conditions are usually compromised to maximize sugar recovery and depending upon what type of pretreatment method is used hemicellulose could be obtained either as a solid fraction or a combination of both solid and liquid fractions (Chandra et al. 2007).

Lignin is the third most abundant polymer in nature. It is a complex, large molecular structure containing cross-linked polymers of phenolic monomers (Fig. 2.3). It is present in plant cell walls and confers a rigid, impermeable resistance to microbial attack and oxidative stress. It is present in the primary cell wall and imparts structural support, impermeability, and resistance against microbial attack. Three phenyl propionic alcohols exist as monomers of lignin. These are:

- Coniferyl alcohol (guaiacyl propanol)
- Coumaryl alcohol (p-hydroxyphenyl propanol)
- Sinapyl alcohol (syringyl alcohol).

Alkyl-aryl, alkyl-alkyl, and aryl-aryl ether bonds link these phenolic monomers together. In general, herbaceous plants such as grasses have the lowest contents of lignin, whereas softwoods have the highest lignin contents (Table 2.1) (Hendricks and Zeeman 2009). Lignin is generally accepted as the 'glue' that binds the different

Table 2.2 Mechanisms of pretreatment

Physical effects
Disrupt the higher order structure
Increase surface area, chemical or enzyme penetration into plant cell walls
Chemical effects
Solubility
Depolymerization
Break crosslinking between macromolecules

components of lignocellulosic biomass together, thus making it insoluble in water. Lignin has been identified as a major deterrent to enzymatic and microbial hydrolysis of lignocellulosic biomass because of its close association with cellulose microfibrils (Avgerinos and Wang 1983). Chang and Holtzapple (2000) showed that biomass digestibility is increased with increasing lignin removal. In addition to being a physical barrier, the harmful effects of lignin include:

- Nonspecific adsorption of hydrolytic enzymes to "sticky" lignin;
- Interference with, and non-productive binding of cellulolytic enzymes to lignin-carbohydrates complexes;
- Toxicity of lignin derivatives to microorganisms.

Different feedstocks contain different amount of lignin that must be removed via pretreatment to enhance biomass digestibility. The lignin is believed to melt during pretreatment and coalesces upon cooling such that its properties are altered; it can subsequently be precipitated (Brownell and Saddler 1987; Converse 1993; Lynd et al. 2002). Delignification (extraction of lignin by chemicals) causes the following effect:

- Biomass swelling
- Disruption of lignin structure
- Increases in internal surface area
- Increased accessibility of cellulolytic enzymes to cellulose fibers.

Although not all pretreatments result in substantial delignification, the structure of lignin may be altered without extraction due to changes in the chemical properties of the lignin. The pretreated biomass becomes more digestible in comparison to the raw biomass even though it may have approximately the same lignin content as non-pretreated biomass. Table 2.2 shows the mechanism of pretreatment.

References

Avgerinos GC, Wang DIC (1983) Selective delignification for fermentation of enhancement. Biotechnol Bioeng 25:67–83

Balaban M, Ucar G (1999) The effect of the duration of alkali pretreatment on the solubility of polyoses. Turk J Agric For 23:667–671

Brownell HH, Saddler JN (1987) Steam pretreatment of lignocellulose materials for enhanced enzymatic hydrolysis. Biotechnol Bioeng 29:228–235
Carere CR, Sparling R, Cicek N, Levin DB (2008) Third generation biofuels via direct cellulose fermentation. Int J Mol Sci 9:1342–1360
Chandra RP, Bura R, Mabee WE, Berlin A, Pan X, Saddler JN (2007) Substrate pretreatment: the key to effective enzymatic hydrolysis of lignocellulosics? Adv Biochem Eng Biotechnol 108:67–93
Chang VS, Holtzapple MT (2000) Fundamental factors affecting biomass enzymatic reactivity. Appl Biochem Biotechnol 84–86:5–37
Converse AO (1993) Substrate factors limiting enzymatic hydrolysis. In: Saddler JN (ed) Bioconversion of forest and agricultural plant residues. CAB International, Wallinfod, pp 93–106
Desvaux M (2005) Clostridium cellulolyticum: model organism of mesophilic cellulolytic clostridia. FEMS Microbiol Rev 29:741–764
Fengel D, Wegener G (1984) Wood chemistry, ultrastructure, Reactions. Walter de Gruyter, Berlin
Hendricks AT, Zeeman G (2009) Pretreatments to enhance the digestibility of lignocellulosic biomass. Bioresour Technol 100:10–108
Laureano-Perez L, Teymouri F, Alizadeh H, Dale BE (2005) Understanding factors that limit enzymatic hydrolysis of biomass: characterization of pretreated corn stover. Appl Biochem Biotechnol 121–124:1081–1099
Levan SL, Ross RJ, Winandy JE (1990) Effects of fire retardant chemical bending properties of wood at elevated temperatures. Research paper FPF-RP-498 Madison, WI: USDA, Forest service; 24
Lynd LR, Weimer PJ, van Zyl WH, Pretorius IS (2002) Microbial cellulose utilization: fundamentals and biotechnology. Microbiol Mol Biol Rev 66:506–577
Mckendry P (2002) Energy production from biomass (part 1) overview of biomass. Bioresour Technol 83:37–46
Palmqvist E, Hahn-Hägerdal B (2000a) Fermentation of lignocellulosic hydrolysates I: inhibition and detoxification. Bioresour Technol 74:17–24
Palmqvist E, Hahn-Hägerdal B (2000b) Fermentation of lignocellulosic hydrolysates II: inhibitors and mechanisms of inhibition. Bioresour Technol 74:25–33
Saha BC (2003) Hemicellulose bioconversion. J Ind Microbiol Biotechnol 2003(30):279–2791
Walker GM (2010) Bioethanol: science and technology of fuel alcohol. Ventus Publishing ApS. ISBN 978-87-7681-681-0

Chapter 3
Conversion of Biomass to Fuel

Abstract The overview on conversion of biomass to fuel is presented in this chapter. The conversion includes the hydrolysis of various components in the lignocellulosic materials to fermentable reducing sugars and the fermentation of the sugars to fuels such as ethanol and butanol. The pretreatment step is mainly required for efficient hydrolysis of cellulose to its constituent sugars. The hydrolysis is usually catalyzed by acids or cellulase enzymes, and the fermentation is carried out by yeasts or bacteria. The factors affecting the hydrolysis of cellulose include porosity of the biomass materials, cellulose fiber crystallinity, and content of both lignin and hemicelluloses.

Keywords Biomass · Fuel · Reducing sugar · Fermentation · Ethanol · Butanol · Pretreatment · Hydrolysis · Cellulose · Enzymes · Porosity · Crystallinity · Lignin · Hemicelluloses

The history of ethanol as a fuel dates back to the early days of the automobile era. However, cheap petrol (gasoline) rapidly replaced ethanol as the fuel of choice, and it was during the late 1970s, when the Brazilian government launched their "Proalcool" Programme, that ethanol made a comeback to the market place. Today, fuel ethanol accounts for roughly two-thirds of world's ethanol production (Saxena et al. 2009). Ethanol is a comparative cleaner burning fuel with high octane and fuel-extending properties. Although blending of ethanol with petrol enhances the volatility of the mixture, on the contrary, ethanol reduces the emission of carbon monoxide from vehicles. The use of petrol blended with 20–24 % ethanol is a standard practice in Brazil. Global ethanol production is shown in Table 3.1. Growth in world ethanol production crucially depends on the development of the fuel alcohol market. Spurred on by the Brazilian "Proalcool" Programme and the US Gasohol scheme, output jumped in the early 1980s, and growth continued at very strong rates up to the mid-1990s. In 1998, fuel alcohol production fell sharply due to the crisis in the Brazilian alcohol sector, which was not compensated by the record output in the United States. Requirement of ethanol in the first phase of the programs on 5 % blending in petrol in India, was 3.45 billion liters a year, which

P. Bajpai, *Pretreatment of Lignocellulosic Biomass for Biofuel Production*,
SpringerBriefs in Green Chemistry for Sustainability,
DOI 10.1007/978-981-10-0687-6_3

Table 3.1 World fuel ethanol production by country or region (million gallons) in 2014

United States	14,300
Brazil	6190
Europe	1445
China	635
Canada	510
Thailand	310
Argentina	160
India	155
Rest of World	865

could have gone up to 5.00 billion liters had the program been introduced throughout the country (Saxena et al. 2009). Several government laboratories, academic institutions, and private sector companies have developed several techniques to accomplish each of the steps required to process the biomass to ethanol. There are different technological options available for the techno-economically feasible process for ethanol production from biomass. Currently, the production of ethanol by fermentation of corn-derived carbohydrates is the major technology used to produce liquid fuels from biomass resources. Dilute acid can open up the biomass structure for subsequent processing. The simultaneous saccharification and fermentation (SSF) process are favored for producing ethanol from the major fraction of lignocellulosic biomass, cellulose, because of its low cost potential.

Figure 3.1 shows a schematic for the conversion of biomass to fuel. The conversion includes the hydrolysis of various components in the lignocellulosic materials to reducing sugars and the fermentation of the reducing sugars to fuels such as ethanol and butanol. The pretreatment step is mainly required for efficient hydrolysis of cellulose to its constituent sugars. The hydrolysis is usually catalyzed by acids or cellulase enzymes, and the fermentation is carried out by yeasts or bacteria. The factors affecting the hydrolysis of cellulose include the following (McMillan 1994):

- Porosity (accessible surface area) of the biomass materials
- Cellulose fiber crystallinity
- Lignin and hemicellulose.

The presence of lignin and hemicellulose makes the accessibility of cellulase enzymes and acids to cellulose more difficult, thus reducing the efficiency of the hydrolysis process. Pretreatment is required to change the size and structure of the biomass and also its chemical composition, so that the hydrolysis of the carbohydrate fraction to monomeric sugars can be obtained rapidly and with greater yields. The hydrolysis process can be improved substantially by removal of lignin and hemicellulose, increase of porosity, and reduction of cellulose crystallinity through pretreatment processes (McMillan 1994). In the hydrolysis process, the sugars are released by breaking down the carbohydrate chains, before they are fermented for alcohol production. The cellulose hydrolysis processes include acid hydrolysis and

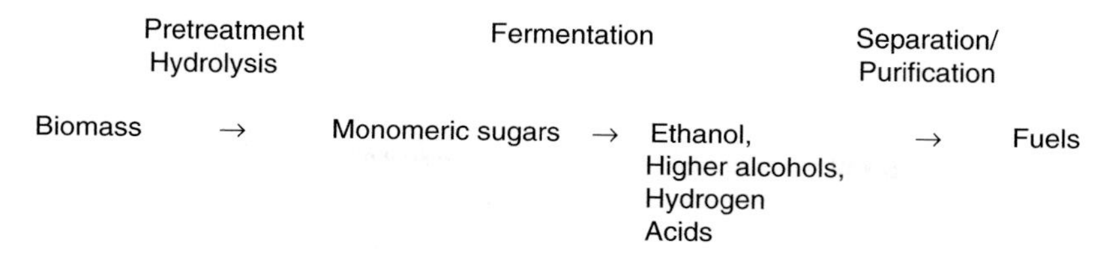

Fig. 3.1 Schematic of the conversion of lignocellulosic biomass to biofuels

an enzymatic hydrolysis. In traditional methods, hydrolysis is performed by reacting the cellulose with an acid. Dilute acid can be used under conditions of both high temperature and pressure whereas concentrated acid can be used at lower temperatures and atmospheric pressure. The decrystallized cellulosic mixture of acid and sugars reacts in the presence of water to liberate individual sugar molecules. The dilute acid process is actually a harsh process that leads to the formation of toxic degradation products that can interfere with fermentation. Cellulose chains can also be broken down into individual glucose sugar molecules by cellulase enzymes. Cellulase refers to a class of enzymes that catalyze the hydrolysis of cellulose. These enzymes are produced mainly by fungi, bacteria, and protozoans. Some cellulases are also produced by plants and animals. The reaction occurs at body temperature in the stomachs of ruminants such as cows and sheep, where the enzymes are produced by intestinal bacteria. Lignocellulosic materials can similarly be enzymatically hydrolyzed under relatively mild conditions. This enables effective cellulose breakdown without the formation of byproducts that would otherwise inhibit enzyme activity. The six-carbon sugars are readily fermented to ethanol by several naturally occurring organisms (Mosier et al. 2005). Traditionally, baker's yeast has been used in the brewing industry to produce ethanol from six-carbon sugars. Due to the complex nature of the carbohydrates present in lignocellulosic biomasses, five-carbon sugars derived from the hemicellulose portion of the lignocellulose are also present in the hydrolysate. For example, the hydrolysate of corn stover contains approximately 30 % of the total fermentable sugars as xylose. As a result, the ability of the fermenting microorganisms to use the whole range of sugars available from the hydrolysate is very much important for increasing the economic competitiveness of cellulosic ethanol and potentially bio-based chemicals. The metabolic engineering of microorganisms used in fuel ethanol production has shown significant progress (Jeffries et al. 2004). For cellulosic ethanol production, several microorganisms such as *Zymomonas mobilis* and *Escherichia coli*, in addition to *Saccharomyces cerevisiae*, have been targeted through metabolic engineering. Several engineered yeasts have been found to efficiently ferment xylose and arabinose and also mixtures of xylose and arabinose (Ohgren et al. 2006; Becker and Boles 2003; Karhumaa et al. 2006). The recovery of fuels from the fermentation broth is obtained by distillation or using a combination of distillation and adsorption. The other components including residual lignin, unreacted cellulose and hemicellulose, and enzymes, accumulate at the bottom of the distillation column.

References

Becker J, Boles E (2003) A modified Saccharomyces cerevisiae strain that consumes L arabinose and produces ethanol. Appl Environ Microbiol 69(7):4144–4150

Jeffries TW, Jin YS (2004) Metabolic engineering for improved fermentation of pentoses by yeasts. Appl Microbiol Biotechnol 63:495–509

Karhumaa K, Wiedemann B, Hahn-Hagerdal B, Boles E, Gorwa-Grauslund MF (2006) Co-utilization of L-arabinose and D-xylose by laboratory and industrial Saccharomyces cerevisiae strains. Microb Cell Fact 10:5–18

McMillan, JD (1994) Pretreatment of lignocellulosic biomass. In: Himmel ME, Baker JO, Overend RP (eds) Enzymatic conversion of Biomass for Fuels production. American Chemical Society, Washington, pp 292–324

Mosier N, Wyman CE, Dale BE, Elander R, Lee YY, Holtzapple MT (2005) Features of promising technologies for pretreatment of lignocellulosic biomass. Bioresour Technol 96:673–686

Ohgren K, Bengtsson O, Gorwa-Grauslund MF, Galbe M, Hahn-Hagerdal B, Zacchi G (2006) Simultaneous saccharification and cofermentation of glucose and xylose in steam-pretreated corn stover at high fiber content with Saccharomyces cereVisiae TMB3400. J Biotechnol 126 (4):488–498

Saxena RC, Adhikari DK, Goyal HB (2009) Biomass-based energy fuel through biochemical routes: a review. Renew Sustain Energy Rev 13:167–178

Chapter 4
Pretreatment of Lignocellulosic Biomass

Abstract Biofuels produced from various lignocellulosic materials have the potential to be a valuable substitute for, or complement to, gasoline. Many physicochemical structural and compositional factors hinder the hydrolysis of cellulose present in biomass to sugars and other organic compounds that can later be converted to fuels. The goal of pretreatment is to make the cellulose accessible to hydrolysis for conversion to fuels. Various pretreatment techniques change the physical and chemical structure of the lignocellulosic biomass and improve hydrolysis rates. A large number of pretreatment methods have been developed. Many methods have been shown to result in high sugar yields. In this chapter, various pretreatment process methods for pretreatment of various lignocellulosic biomasses are presented.

Keywords Pretreatment techniques · High-energy radiation · Physicochemical pretreatment · Steam explosion · Pyrolysis · Liquid hot water pretreatment · Ammonia fiber/freeze explosion · Ammonia recycle percolation · Soaking aqueous ammonia · Carbon dioxide explosion · Chemical pretreatment · Ozonolysis · Acid treatment · Alkali treatment · Oxidative delignification · Hydrogen peroxide · Organosolv process · Sulfite pretreatment · Ionic liquids · Biological pretreatment · Wet oxidation · Microwaves pretreatment

The main objectives of the pretreatment process are to remove lignin and hemicelluloses, increase the porosity of the lignocellulosic materials, and reduce the crystallinity of cellulose (Fig. 4.1). Pretreatment should meet the following requirements:

- Low capital and operational cost
- Effective on a wide range and loading of lignocellulosic materials
- Should result in the recovery of most of the lignocellulosic components in a useable form in separate fractions
- Need for preparation/handling or preconditioning steps prior to pretreatment such as size reduction should be minimized
- Avoid the formation of byproducts that are inhibitory to the subsequent hydrolysis and fermentation processes

P. Bajpai, *Pretreatment of Lignocellulosic Biomass for Biofuel Production*,
SpringerBriefs in Green Chemistry for Sustainability,
DOI 10.1007/978-981-10-0687-6_4

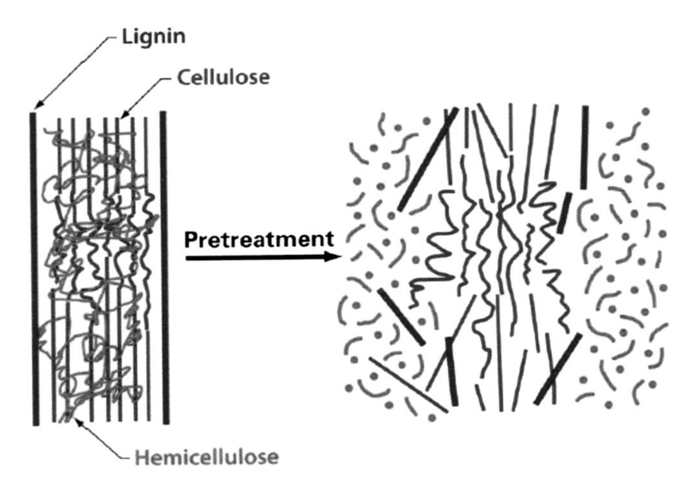

Fig. 4.1 Schematic of pretreatment effect on lignocellulosic biomass. Based on Mosier et al. (2005a, b)

– Low energy demand or be performed in a manner that energy invested could be used for other purposes such as secondary heating.
– Improve the formation of sugars or the ability to subsequently form sugars by hydrolysis,
– Avoid the degradation or loss of carbohydrates.

Other features which form the basis of comparison of different pretreatment options include the following (Chandra et al. 2007; Galbe and Zacchi 2007; Mosier et al. 2005b; Lynd et al. 1996, 2003, 2008):

– Regeneration/cost of catalyst
– Generation of higher value lignin coproducts
– Obtaining hemicellulose sugars in the liquid phase to reduce the need for the use of hemicellulases in subsequent enzymatic hydrolysis.

Pretreatments employed can be divided into physical, chemical, and biological methods but there is a strong interdependence of these processes. There is not a perfect pretreatment method employed and remaining bottlenecks include formation of inhibitory products such as acids, furans, phenols, high particle load, high energy input, and efficient separation of soluble sugars from solid residues. Specific pretreatment conditions are needed for individual feedstocks and mechanistic models can help in the rational design of such processes (Zhang 2008; Zhang et al. 2009; Eggeman and Elander 2005). It is particularly important to optimize lignocellulose pretreatment methods because they are one of the most expensive steps in the overall conversion to bioethanol. Mosier et al. (2005b) reported that pretreatment accounts for ~30 US cents/gallon of cellulosic ethanol produced. Pretreatment is needed to change the biomass macroscopic and microscopic size and structure and also submicroscopic chemical composition and structure so that hydrolysis of carbohydrate fraction to monomeric sugars can be obtained more rapidly and with higher yields (Sun and Cheng 2002; Mosier et al. 2005a, b; Tucker et al. 2003). Pretreatment of cellulosic biomass in cost-effective manner is a major challenge of cellulose to ethanol technology research and development. Native lignocellulosic biomass is extremely recalcitrant to enzymatic digestion. Therefore, a number of thermochemical pretreatment methods have been developed to improve digestibility (Wyman et al. 2005a, b). Several studies have clearly shown that there is a direct correlation between the removal of lignin and hemicellulose on cellulose digestibility (Kim and Holtzapple 2006a, b). Thermochemical processing options are found to be more promising than biological options for the conversion of lignin fraction of cellulosic biomass, which can have an adverse effect on enzyme hydrolysis. It can also serve as a source of process energy and potential coproducts that have important benefits in a life cycle context (Sheehan et al. 2003). Pretreatment can be carried out using different methods such as mechanical combination, steam explosion, ammonia fiber explosion, acid or alkaline pretreatment and biological treatment, organosolv pretreatment etc. (Cadoche and López 1989; Gregg and Saddler 1996; Kim et al. 2003; Damaso et al. 2004; Kuhad et al. 1997; Keller et al. 2003; Itoh et al. 2003; Kuo and Lee 2009b; Zhao and Liu 2012; Zhang et al. 2009, 2012, 2013a; Brink 1994; Palmqvist and Hahn-Hägerdal 2000b; Takacs et al. 2000; Ang et al. 2012; Wyman et al. 2009).

To assess the cost and performance of pretreatment methods, technoeconomic analysis has been conducted (Eggeman and Elander 2005).

4.1 Pretreatment Methods

Pretreatment methods can be divided into different categories such as physical, physicochemical, chemical, biological, electrical, or a combination of these. In combinatorial pretreatment methods, physical parameters such as pressure or temperature or a biological step are combined with chemical treatments and are termed physicochemical or biochemical pretreatment methods. Ammonia fiber/freeze explosion (AFEX) is a good example of a physicochemical method (Sun and Cheng 2002), and bioorganosolv is a good example of a biochemical method for biomass pretreatment (Itoh et al. 2003). Combinatorial pretreatment methods are generally found to be more effective in improving the biomass digestibility, and are often used in designing leading pretreatment technologies.

4.1.1 Physical Pretreatment

4.1.1.1 Mechanical Comminution

Mechanical comminution increase the available specific surface area, and reduce degree of polymerization (DP) and also the cellulose crystallinity (Sun and Cheng

2002). Different mechanical size-reduction methods have been used to improve the digestibility of lignocellulosic biomass (Palmowski and Muller 1999). These are coarse size reduction, chipping, shredding, grinding and milling (hammer- and ball-milling [wet, dry, vibratory rod/ball milling]), compression milling, ball milling/beating, agitation bead milling, and pan milling. Other types of milling such as fluid energy milling, two-roll milling, and colloid milling have been also used (Murnen et al. 2007; Isci et al. 2008; Kim and Lee 2005a, b; Dale and Moreira 1982; Ryu and Lee 1982). Attrition and disk refining have been also explored for pretreatment (Mes-Hartree et al. 1988; Murnen et al. 2007). Biomass has been also pretreated by simultaneous ball milling/attrition and enzymatic hydrolysis (Zheng et al. 1998; Ben-Ghedalia and Miron 1981). Vibratory ball milling has been found to be more effective than ordinary ball milling for improving the biomass digestibility when used to pretreat aspen and spruce chips (Neely 1984). Only compression milling process has been studied in production scale (Kilzer and Broido 1965).

The objective of chipping is to reduce heat and mass transfer limitations. Grinding and milling are more effective at reducing the particle size and cellulose crystallinity than chipping probably as result of the shear forces generated during milling. The type and duration of milling and also the kind of biomass determine the increase in specific surface area, final degree of polymerization, and the net reduction in cellulose crystallinity. Vibratory ball milling is more effective than ordinary ball milling in reducing cellulose crystallinity of spruce and aspen chips. Disk milling which produces fibers is more effective in enhancing cellulose hydrolysis than hammer milling which produces finer bundles (Zhua et al. 2009). Chipping reduces the biomass size to 10–30 mm while grinding and milling can reduce the particle size to 0.2–2 mm. Further reduction of biomass particle size below 40 mesh (0.400 mm) has little effect on the rates and yields of biomass hydrolysis (Chang et al. 1997).

The energy requirements of mechanical comminution of lignocellulosic biomass depend on the following factors:

- Characteristics of biomass
- Final particle size required.

Compared to agricultural residues, more energy is required in case of hardwoods (Cadoche and López 1989). Size reduction has been used in most studies of hydrolysis. But not much information is available on the characteristics of the substrate and also the energy consumed during the process (Zhu et al. 2005a, b). Studies have shown that milling process increases bioethanol, biogas, and biohydrogen yields (Delgenes et al. 2002). Taking into consideration the high energy requirement of milling on an industrial scale and the increase in energy demands, it is unlikely that milling is still economically feasible (Hendricks and Zeeman 2009). Milling can be performed before or after the chemical pretreatment. Studies have shown that performing milling after chemical pretreatment results in the following benefits (Zhu et al. 2010a; Zhua et al. 2009):

- Significantly reduces milling energy consumption
- Does not result in the generation of fermentation inhibitors
- Reduces cost of solid liquid separation because the pretreated chips can be easily separated
- Eliminates energy-intensive mixing of pretreatment slurries and liquid to solid ratio.

4.1.1.2 High-Energy Radiation

Digestibility of cellulosic biomass can be improved by using the high-energy radiation methods. These include the following (Yang et al. 2008; Youssef and Aziz 1999; Imai et al. 2004; Nitayavardhana et al. 2008; Wojciak and Pekarovicova 2001; Bak et al. 2009; Shin and Sung 2008; Dunlap and Chiang 1980; Kitchaiya et al. 2003; Maa et al. 2009; Saha et al. 2008; Zhu et al. 2005a, b; Yang et al. 2008):

- Gamma rays
- Ultrasound
- Electron beam
- Pulsed electrical field
- Ultraviolet rays
- Microwave heating.

The action mode behind the high-energy radiation could be attributed to

- One or more changes of features of cellulosic biomass including increase of specific surface area
- Decrease of the degrees of polymerization and crystallinity of cellulose
- Hydrolysis of hemicelluloses
- Partial depolymerization of lignin.

However, these high-energy radiation methods are mostly energy-intensive, slow, and expensive (Chang et al. 1981; Lin et al. 1981). They are also strongly substrate-specific (Dunlap and Chiang 1980). Therefore, high-energy radiation techniques lack commercial appeal based on current estimation of overall cost.

4.1.1.3 Pyrolysis

Pyrolysis has also been examined for pretreatment of lignocellulosic feedstocks (Kumar et al. 2009). When the materials are treated at temperatures higher than 300 °C, cellulose rapidly decomposes and produces gaseous products and residual char (Kilzer and Broido 1965; Shafizadeh and Bradbury 1979). The decomposition is much slower and less-volatile products are formed at lower temperatures. Mild acid hydrolysis (1N sulfuric acid, 2.5 h, 97 °C) of the residues from pyrolysis pretreatment has shown 80–85 % conversion of cellulose to reducing sugars with more than 50 % glucose (Fan et al. 1987). The process can be improved in the presence of oxygen (Shafizadeh and Bradbury 1979). The decomposition of pure cellulose can take place at a lower temperature in the presence of catalyst (zinc chloride or sodium carbonate) (Shafizadeh and Lai 1975). Zwart et al. (2006) reported production of transportation fuels from biomass via a so-called biomass-to-liquids route, in which biomass is converted to syngas from which high-quality Fischer–Tropsch fuels are produced. Chipping, pelletization, torrefecation, and pyrolysis have been studied as pretreatment processes for biomass-to-Fischer–Tropsch-fuel conversion. Pretreatment by torrefaction was found to be far more attractive than pyrolysis.

4.1.2 Physicochemical Pretreatment

4.1.2.1 Steam Explosion

Steam explosion is the mostly applied physicochemical method of biomass pretreatment and has been studied extensively (Agbor et al. 2011). The term "autohydrolysis" has also been used as a synonym for steam explosion describing the changes that take place during this process (Chandra et al. 2007; McMillan 1994a, b; Saddler et al. 1993). In this process biomass is first chipped, ground, or simply raw preconditioned and then physically pretreated with high-pressure saturated steam at pressures between 0.7 and 4.8 MPa and temperatures of about 160–240 °C (Agbor et al. 2011). The pressure is maintained for several seconds to a few minutes to promote hydrolysis of hemicelluloses and then released. The process causes degradation of hemicelluloses and lignin transformation due to high temperature, thus increasing the potential of cellulose hydrolysis. Ninety percent efficiency of enzymatic hydrolysis has been obtained in 24 h for poplar chips pretreated by steam explosion, compared to only 15 % hydrolysis of untreated chips (Grous et al. 1986). Several factors have been found to affect steam explosion pretreatment. These are moisture content, temperature, residence time, and the size of the chips (Duff and Murray 1996). Optimal hemicellulose solubilization and hydrolysis can be obtained by using either high temperature and short residence time (270 °C, 1 min) or lower temperature and longer residence time (190 °C, 10 min) (Duff and Murray 1996). Studies have shown that lower temperature and longer residence time are more promising because they avoid the formation of degradation products of sugars that inhibit subsequent fermentation (Wright 1998).

Hemicellulose is the main fraction of the carbohydrates which are solubilized in the liquid phase during pretreatment, whereas the lignin is transformed as result of the high temperature. The cellulose in the solid fraction becomes more accessible; therefore, the digestibility of the lignocellulosic feedstock is increased. Hydrolysis of hemicellulose is thought to be mediated by the acetic acid produced from acetyl groups associated with hemicellulose and other acids liberated during pretreatment, that may further catalyze the hemicellulose hydrolysis resulting in the liberation of monomers of glucose and xylose therefore, the use of the term autohydrolysis (Mosier et al. 2005b; Weil et al. 1997).

Steam explosion can be effectively improved by the addition of, carbon dioxide or sulfur dioxide, sulfuric acid as a catalyst. The use of acid catalyst has been found to result in several benefits such as increasing the recovery of hemicellulose sugars, decreasing the production of inhibitory compounds, and improving the enzymatic hydrolysis on the solid residue (Mosier et al. 2005b; Sun and Cheng 2002). The advantages and disadvantages of steam explosion are presented in Table 4.1.

Steam explosion is found to be very effective for the pretreatment of agricultural residues and hardwoods, but not found to be much effective for softwoods (Agbor et al. 2011). In this case, the use of an acid catalyst is particularly important. This process is approaching commercialization and has been studied in pilot scale at NREL pilot plant in Golden, SEKAB pilot plant in Sweden, the Italian steam explosion programme in Trisaia southern Italy, and by a demonstration-scale ethanol plant at Iogen in Ottawa, Canada. The uncatalyzed steam pretreatment process has been used commercially in the masonite process for the production of fiber board and other products (Avella and Scoditti 1998; Galbe and Zacchi 2007; Mosier et al. 2005a, b).

4.1.2.2 Liquid Hot Water Pretreatment

Liquid hot water pretreatment (LHW) process is similar to steam explosion but uses water in the liquid state at high temperatures instead of steam (Agbor et al. 2011). Following terms have been used to describe LHW for pretreatment of biomass:

Table 4.1 Advantages and disadvantages of steam explosion

Advantages
Makes limited use of chemicals
Does not result in excessive dilution of the resulting sugars
Requires low energy input with no recycling or environmental cost
Disadvantages
Incomplete destruction of lignin–carbohydrate matrix resulting in the risk of condensation and precipitation of soluble lignin components making the biomass less digestible
Destruction of a portion of the xylan in hemicellulose and possible generation of fermentation inhibitors at higher temperatures
Because of the formation of degradation products that are inhibitory to microbial growth, enzymatic hydrolysis, and fermentation, pretreated biomass needs to be washed by water to remove the inhibitory materials along with water-soluble hemicellulose. This decreases the overall saccharification yields due to the removal of soluble sugars, such as those generated by hydrolysis of hemicellulose

Based on McMillan (1994a, b), Mes-Hartree et al. (1988)

- Solvolysis
- Hydrothermolysis
- Aqueous fractionation
- Aquasolv.

This process results in lignin removal and hemicellulose hydrolysis and renders cellulose in the biomass more accessible (Yang and Wyman 2004). It avoids the formation of fermentation inhibitors that occur at higher temperatures. In this process, lower temperatures—optimum between 180 and 190 °C for corn stover and low dry matter—about 1–8 % content are used resulting in the production of more poly and oligosaccharides. The temperature of 160–190 °C for pH-controlled LHW pretreatment and 170–230 °C have been reported depending on the severity of the pretreatment (Wyman et al. 2005a, b; Bobleter 1994). LHW pretreatment has been conducted in three different types of reactor configurations (cocurrent, countercurrent and flow through reactor configuration) depending on the direction of the flow of the water and biomass into the reactor. Water and biomass are brought in contact at temperatures of 200–230 °C for up to 15 min. Hot water breaks the hemiacetal linkages and liberates acids during biomass hydrolysis. This facilitates the breakage of ether linkages in biomass (Antal 1996). Mosier et al. (2005b) stated that the cleavage of O-acetyl groups and uronic acid substitutions on the hemicellulose could help or interfere LHW pretreatment, because the release of these acids helps to catalyze the formation and removal of oligosaccharides, or further hydrolyze hemicellulose to monomeric sugars, which can be subsequently degraded to aldehydes i.e., furfural from pentoses and 5-hydroxymethyl furfural from hexoses which inhibit microbial fermentation (Palmqvist and Hahn-Hägerdal 2000a, b). The production of monosaccharides and the subsequent degradation products that further catalyze cellulosic hydrolysis during LHW pretreatment can be reduced by maintaining the pH between 4 and 7 (Kohlmann et al. 1995). The flow through reactor configuration in which hot water is passed over a stationary bed of lignocellulose was reported by Yang and Wyman (2004). It was found to be the more effective configuration for removing hemicellulose and lignin at same severity. High-lignin solubilization hinders recovery of hemicelluloses (Mok and Antal 1992, 1994). Acid catalyst can be added making the process similar to dilute acid pretreatment. But catalytic degradation of sugars results in undesirable side products. During pretreatment, the pKa of water and pH are affected by temperature; so, potassium hydroxide is used to maintain the pH above 5 and below 7 to reduce the formation of monosaccharides that are degraded to fermentation inhibitors (Mosier et al. 2005b; Weil et al. 1998).

LHW pretreatment has been studied at lab scale. pH-controlled LHW pretreatment is considered for large-scale pretreatment of corn fiber. Mosier et al. (2005a) have reported successful pretreatment of corn fiber slurry in a 163 L/min reactor with a 20-min residence time, showing the possibility of scaling-up LHW pretreatment and its application in pretreating large quantities of corn fiber.

Table 4.2 shows the advantages and disadvantages of steam explosion.

Table 4.2 Advantages and disadvantages of LHW

Advantages
Formation of degradation products is minimized due to the use of lower temperatures. This eliminates the need for a final washing step or neutralization because the pretreatment solvent here is water
Low cost of the solvent is also an advantage for large-scale application
Disadvantages
The amount of solubilized product is higher, while the concentration of these products is lower compared to steam explosion (Bobleter 1994)
More energy demanding because of the large volumes of water involved

Based on Bobleter (1994)

4.1.2.3 Ammonia Fiber/Freeze Explosion, Ammonia Recycle Percolation and Soaking Aqueous Ammonia

Ammonia fiber/freeze explosion (AFEX), Ammonia recycle percolation (ARP), and Soaking aqueous ammonia (SAA) are alkaline pretreatment methods that use liquid ammonia to pretreat the biomass (Agbor et al. 2011; Kumar et al. 2009). In AFEX process, lignocellulosic materials are exposed to liquid ammonia at high temperature and pressure for a period of time, and then the pressure is swiftly reduced. In ARP process, aqueous ammonia passes through biomass at high temperature, after which ammonia is recovered. As a physicochemical process, AFEX is similar to steam explosion operating at high pressure but it is performed at ambient temperatures, while ARP is carried out at higher temperatures. SAA is a modified version of AFEX utilizing aqueous ammonia to treat biomass in a batch reactor at moderate temperatures to reduce the liquid throughput during pretreatment (Kim and Lee 2005a, b). At ambient temperatures, the duration could be up to 10–60 days while at higher temperatures, the effect of ammonia is rapid and the duration of pretreatment is reduced to minutes (Alizadeh et al. 2005; Kim et al. 2000).

The concept of AFEX is similar to steam explosion. AFEX pretreatment can substantially improve the saccharification rates of various herbaceous crops and grasses. It can be used for the pretreatment of several lignocellulosic materials such as alfalfa, wheat straw, wheat chaff (Mes-Hartree et al. 1988), barley straw, corn stover, rice straw (Vlasenko et al. 1997), bagasse (Holtzapple et al. 1991), coastal Bermuda grass, switchgrass (Reshamwala et al. 1995), municipal solid waste, softwood newspaper, kenaf newspaper (Holtzapple et al. 1992a), and aspen chips (Tengerdy and Nagy 1988). Compared to acid pretreatment and acid-catalyzed steam explosion, the AFEX pretreatment does not considerably solubilize hemicellulose (Mes-Hartree et al. 1988; Vlasenko et al. 1997). Mes-Hartree et al. (1988) made a comparison of the steam and ammonia pretreatment for enzymatic hydrolysis of wheat straw, wheat chaff, alfalfa stems and aspen wood. They observed that steam explosion solubilized the hemicellulose, while AFEX did not.

After AFEX pretreatment, the composition of the materials was found to be the same as the original materials. Over 90 % hydrolysis of cellulose and hemicellulose was obtained after AFEX pretreatment of Bermuda grass (approximately 5 % lignin) and bagasse (15 % lignin) (Holtzapple et al. 1991). However, the AFEX process was not very effective for the biomass with high lignin content such as newspaper (18–30 % lignin) and aspen chips (25 % lignin). Hydrolysis yield of AFEX-pretreated newspaper and aspen chips was reported as only 40 % and below 50 %, respectively (McMillan 1994a, b). Ammonia must be recycled after the pretreatment in order to reduce the cost and protect the environment. In an ammonia recovery process, a superheated ammonia vapor with a temperature up to 200 °C was used to vaporize and strip the residual ammonia in the pretreated biomass and the evaporated ammonia was then withdrawn from the system by a pressure controller for recovery (Holtzapple et al. 1992b). The ammonia pretreatment does not produce inhibitors for the downstream biological processes, so water wash is not required (Dale et al. 1984; Mes-Hartree et al. 1988). Holtzapple et al. (1990) reported that AFEX pretreatment does not require small particle size for efficacy.

In the AFEX and ARP processes, the lignocellulosic material is exposed to ammonia at a given temperature and high pressure. This results in swelling and phase change in cellulose crystallinity of biomass in addition to the alteration and removal of lignin. The reactivity of the remaining carbohydrates after pretreatment is increased. The pretreated biomass is easily hydrolyzable with close to theoretical yields after enzymatic hydrolysis at low enzyme loadings compared to pretreated biomass from other PPMs (Foster et al. 2001; Holtzapple et al. 1991; Kim and Lee 2002; Vlasenko et al. 1997). In AFEX pretreatment, biomass is brought in contact with anhydrous liquid ammonia loading (1–2 kg of ammonia/kg of dry biomass) at 60–90 °C for 10–60 min and pressures above 3 MPa. In a closed vessel under pressure, the biomass and ammonia mix is heated for about 30 min to the desired temperature. After holding the target temperature for about 5 min, the vent valve is opened rapidly to explosively relieve the pressure. The rapid release causes evaporation of the ammonia (which is volatile at atmospheric pressure) and a concomitant drop in temperature of the system (Alizadeh et al. 2005; Dale and Moreira 1982). The chemical effect of ammonia under pressure causes the cellulosic biomass to swell, thus increasing the accessible surface area while decrystallizing cellulose. This results in a phase change in the crystal structure of cellulose I to cellulose III (Mosier et al. 2005b; O'Sullivan 1996). A small amount of hemicellulose is solubilized in the oligomeric form during AFEX pretreatment. The lignin distribution in the biomass remains comparatively the same after AFEX pretreatment, but the lignin structure is rigorously changed. This results in the increase of water-holding capacity and digestibility. The combined physical and chemical changes significantly increase the susceptibility of the pretreated lignocellulosic biomass to subsequent enzymatic hydrolysis (Dale et al. 1984; Dale and Moreira 1982; Galbe and Zacchi 2007; Holtzapple et al. 1991; Kim and Lee 2005a, b). The mild process conditions reduce the formation of sugar degradation products and fermentation inhibitors. AFEX pretreatment can significantly improve the saccharification rates of herbaceous plants, agricultural residues, and municipal solid waste with approximately 90 % of the theoretical hydrolysis of AFEX-pretreated coastal Bermuda grass (Holtzapple et al. 1992a; Mes-hartree et al. 1988).

ARP uses ammonia percolating in a flowthrough mode through a packed bed reactor (Agbor et al. 2011). In this process, aqueous ammonia (5–15 %) passes through biomass in a flowthrough column reactor at high temperatures (150–180 °C) and a flow rate of 1–5 mL/minutes with residence time of 10–90 min, after which the ammonia is recycled or recovered (Kim et al. 2003). There is simultaneous lignin removal and hydrolysis of hemicellulose under these conditions and cellulose crystallinity is reduced. Kim et al. (2006), reported a low-liquid ARP using one reactor void volume with 15 wt% ammonia in an attempt to optimize the process and establish a continuous process, by dropping the liquid ammonia throughput. The digestibility of the low-liquid ARP-pretreated biomass under optimum conditions was at least 86 % of the theoretical maximum even at the lowest enzyme loading of 7.5 FPU/g glucan. Both AFEX and ARP are effective pretreatments for herbaceous plants, agricultural residues and municipal solid waste. ARP pretreatment is effective on hardwoods also (Iyer et al. 1996; Kim and Lee 2005a, b).

Chan et al. (2014) performed a carbon dioxide-added ammonia explosion pretreatment for bioethanol production from rice straw. Using response surface methodology, the pretreatment conditions, such as ammonia concentration, carbon dioxide loading level, residence time, and temperature were optimized. The response for optimization was defined as the glucose conversion rate. The optimized pretreatment conditions resulting in maximal glucose yield (93.6 %) were determined as 14.3 % of ammonia concentration, 2.2 MPa of carbon dioxide loading level, 165.1 °C of temperature, and 69.8 min of residence time. Scanning electron microscopy analysis showed that pretreatment of rice straw strongly increased the surface area and pore size, which increased enzymatic accessibility for enzymatic saccharification. Finally, an ethanol yield of 97 % was obtained via simultaneous saccharification and fermentation suggesting that carbon dioxide added ammonia pretreatment is an appropriate process for bioethanol production from rice straw.

Table 4.3 shows the advantages and disadvantages of AFEX/ARP (Agbor et al. 2011). These processes have not been reported beyond lab-scale use. It is possible, however, that AFEX- and ARP-pretreated biomass could be used as feed for cattle serving as a source of essential nutrients. This is a potential application of AFEX/ARP to produce value-added products for IFB.

4.1.2.4 Carbon Dioxide Explosion

Carbon dioxide explosion process involves the use of supercritical carbon dioxide under pressure to enhance the digestibility of lignocellulosic biomass (Agbor et al. 2011). Supercritical carbon dioxide explosion, has a lower temperature than steam explosion and possibly has a reduced expense compared to ammonia explosion. Supercritical fluid refers to a fluid that is in a gaseous form but is compressed at

Table 4.3 Advantages and disadvantages of AFEX/ARP

Advantages
Hydrolyzate from AFEX/ARP is compatible with fermentation organisms without the need for conditioning
Ability to reduce, recover, and recycle the ammonia used in both AFEX/ARP makes the establishment of a continuous industrial process more feasible
Moderate temperatures, pH, and short residence time used in AFEX mostly and ARP compared to other physicochemical processes minimizes the formation of fermentation inhibitors
High selectivity for reaction with lignin
ARP pretreatment is an efficient and selective delignification method reducing 70–85 % of corn stover lignin within 20 min of pretreatment and removing 40–60 % hemicellulose while leaving the cellulose intact
Need for water washing step is eliminated
Total sugar yield is increased
AFEX/ARP pretreatment is cheaper than DAP
Disadvantages
Somewhat ineffective in the pretreatment of high-lignin-containing lignocellulosic biomass
Cost of ammonia basically drives the process and its application on large scale
Environmental concerns with the stench of ammonia also have a negative impact on pilot and industrial scale applications

temperatures above its critical point to a liquid like density. High-pressure carbon dioxide, and particularly supercritical carbon dioxide is injected into the reactor and then liberated by an explosive decompression. Because carbon dioxide forms carbonic acid when dissolved in water, the acid increases the hydrolysis rate. Carbon dioxide molecules are comparable in size to water and ammonia and should be able to penetrate small pores accessible to water and ammonia molecules. Carbon dioxide was suggested to be helpful in hydrolyzing hemicellulose and also cellulose. Moreover, the low temperature prevents any appreciable decomposition of monosaccharides by the acid. Upon an explosive release of the carbon dioxide pressure, the disruption of the cellulosic structure increases the accessible surface area of the substrate to hydrolysis.

The supercritical carbon dioxide pretreatment method does not discharge harmful chemicals (Agbor et al. 2011). So it is an environment friendly method with its own niche in tactical biomass processing. Research on its use has been conducted on few lignocellulosic materials such as Avicel cellulose, recycled paper, sugarcane bagasse, aspen, Southern yellow pine, corn stover, switch grass (Zheng et al. 1998; Kim and Hong 2001; Narayanaswamy et al. 2011).

In this process, supercritical carbon dioxide is delivered to biomass placed in a high-pressure vessel (Kim and Hong 2001), or delivered at high pressure (1000–4000 psi) (Zheng et al. 1995). The vessel is heated to the desired temperature and held for a designated length of time, usually several minutes when conducted at high temperatures up to 200 °C (Hendricks and Zeeman 2009). Carbon dioxide penetrates the biomass at high pressure and it is believed that once dissolved in water, carbon dioxide will produce carbonic acid. This acid helps in the hydrolysis of hemicellulose. The release of the pressurized gas results in the disruption of the biomass native structure increasing the accessible surface area (Zheng et al. 1995). Carbon dioxide explosion pretreatment is not effective on biomass with no moisture content but increases the hydrolytic yield of moisture containing biomass with the yield increasing proportionately with the moisture content of the unprocessed feedstock (Kim and Hong 2001). The attractive features of carbon dioxide explosion pretreatment are presented below:

- Low cost
- No generation of toxins
- Use of low temperatures
- High solids capacity.

However, the high cost of equipment that can withstand high pressure conditions of carbon dioxide explosion pretreatment is a strong limitation to the application of this process on a large scale. Furthermore, the effects on biomass carbohydrate components have yet to be elucidated.

4.1.3 Chemical Pretreatment

4.1.3.1 Oxidative Delignification

Oxidizing agent such as hydrogen peroxide, ozone, oxygen or air can also be used for delignification of lignocellulosic feedstocks (Bensah and Mensah 2013; Hammel et al. 2002; Nakamura et al. 2004). In this process, lignin is converted to acids. However, these acids can act as inhibitors during fermentation process. Hence, these acids have to be removed (Alvira et al. 2010). Also, oxidative treatment damages the hemicellulose fraction of the lignocellulose complex and a major portion of the hemicellulose gets degraded and becomes unavailable for fermentation (Lucas et al. 2012). The major types of oxidative delignification process are as follows.

Hydrogen Peroxide

Hydrogen peroxide is the most commonly used oxidizing agent. It degrades into hydrogen and oxygen and does not leave residues in the biomass (Uppal et al. 2011). Process variations that have also produced high sugar yields include the addition of catalysts such as manganese acetate, post-pretreatment acid saccharification, and alkaline-peroxide application without post-pretreatment washing (Lucas et al. 2012; Uppal et al. 2011; Banerjee et al. 2012). By pretreating water hyacinth and lettuce with sodium hydroxide followed by hydrogen peroxide, Mishima et al. (2006) obtained higher sugar yields compared to sodium hydroxide, sulfuric acid, and hot water pretreatments under similar conditions. Studies have shown that dissolution of about 50 % of lignin and most of the hemicellulose has been obtained in a solution of 2 % hydrogen peroxide at 30 °C. Hydrolysis of hydrogen peroxide leads to the generation of hydroxyl radicals, which degrade lignin and produce low molecular weight products. Removal of lignin from lignocellulose leads to the exposure of cellulose and hemicellulose causing increased enzymatic hydrolysis (Hammel et al. 2002). The yield of enzymatic hydrolysis followed can be as high as 95 %. Combined chemical (hydrogen peroxide) and biological treatment of rice hull was performed by Yu et al. (2009). Their study consisted of chemical treatment (hydrogen peroxide) followed by biological treatment with *Pleurotus ostreatus*. The combined pretreatments led to significant increases of the lignin degradation. At optimum conditions, hydrogen peroxide (2 %, 48 h) and *P. ostreatus* (18 days) led to 39.8 and 49.6 % of net yields of total sugar and glucose, respectively. It was about 5.8 times and 6.5 times more than that of the fungal pretreatment (18 days) alone, and was also slightly higher as compared to sole fungal pretreatment for 60 days.

A study aimed at increasing the enzymatic digestibility of sweet sorghum bagasse for bioethanol production was performed by Cao et al. (2012). Among different chemical pretreatment methods employed, the best yields were obtained when sweet sorghum bagasse was treated with dilute sodium hydroxide solution followed by autoclaving and hydrogen peroxide immersing pretreatment. The highest cellulose hydrolysis yield, total sugar yield and ethanol concentration were 74.3, 90.9 % and 6.1 g/L, respectively, which were 5.9, 9.5, and 19.1 times higher in comparison to the control.

Lucas et al. (2012) studied the effect of manganese acetate as a catalyst in oxidative delignification of wood with hydrogen peroxide at room temperature. Poplar wood sections were converted into a fine powder-like material when exposed to a mixture of hydrogen peroxide and manganese acetate in aqueous solution. Optical and Raman microscopy showed oxidation of lignin containing middle lamellae. Raman spectra from the solid residue showed a delignified and cellulose-rich material. Glucose yields following enzymatic hydrolysis were 20–40 % higher in poplar sawdust pretreated with hydrogen peroxide and manganese acetate for 2, 4, and 7 days at room temperature as compared to those in sawdust exposed to water only for identical durations, suggesting the viability of the mild, inexpensive method for pretreatment of lignocellulosic biomass.

Alkaline Hydrogen Peroxide pretreatment of Cashew Apple Bagasse was studied by Correia et al. (2013). The effects of the concentration of hydrogen peroxide at pH 11.5, the biomass loading and the pretreatment duration performed at 35 °C and 250 rpm were evaluated after the subsequent enzymatic saccharification of the pretreated biomass using a commercial cellulase enzyme. At optimized conditions, consisting of solid loading of 5 % (w/v) at 4.3 % alkaline hydrogen peroxide, 6 h, 35 °C, a total reducing sugar yield of 42.9 % was obtained.

Pretreatment with hydrogen peroxide is considered a very harsh treatment which leads to lignin removal and high yields of reducing sugars. Reducing sugar yields up to 90 % have been obtained (Cao et al. 2012).

Table 4.4 shows advantages and disadvantages of hydrogen peroxide pretreatment.

Ozonolysis

Ozonolysis can be used to reduce the lignin content of lignocellulosic wastes. This increases the in vitro digestibility of the treated material, and does not produce toxic residues unlike other chemical treatments. Ozone can be used to degrade lignin and hemicellulose in several lignocellulosic materials such as poplar saw dust, wheat straw, bagasse, green hay, peanut, pine and cotton straw (Ben-Ghedalia and Miron 1981; Neely 1984; Vidal and Molinier 1988). The degradation was mainly limited to lignin; hemicellulose was slightly attacked, whereas cellulose was hardly affected. The rate of enzymatic hydrolysis increased by a factor of 5 following 60 % removal of the lignin from wheat straw in ozone pretreatment (Vidal and Molinier 1988). Enzymatic hydrolysis yield increased from 0 to 57 % as the percentage of lignin decreased from 29 to 8 % after ozonolysis pretreatment of poplar sawdust (Vidal and Molinier 1988). Table 4.5 shows the advantages and disadvantages of ozonolysis pretreatment.

Table 4.4 Advantages and disadvantages of hydrogen peroxide pretreatment

Advantages
Hydrogen peroxide permits fractionation of biomass at ambient pressures and low temperatures allowing the use of low-cost reactors
Does not leave any residue in biomass
Disadvantages
Application of oxidizing agents produces soluble lignin compounds that inhibit the conversion of hemicelluloses and cellulose to ethanol
Loss of sugar due to the occurrence of nonselective oxidation
High cost of oxidants is a major limitation for scaling up to industrial levels

Table 4.5 Advantages and disadvantages of ozonolysis pretreatment

Advantages
Effectively removes lignin
Does not produce toxic residues for the downstream processes
Reactions are carried out at room temperature and pressure
Can be easily decomposed by using a catalytic bed or increasing the temperature—processes can be designed to minimize environmental pollution
Disadvantages
Large amount of ozone is required, making the process expensive

Most ozonation experiments have been conducted in hydrated fixed beds which lead to more effective oxidations than aqueous suspension or suspensions in 45 % acetic acid (Vidal and Molinier 1988; Lasry et al. 1990). Morrison and Akin (1990) used ozone to oxidize herbaceous species moistened to 50 %. Caproic, levulinic, p-hydroxybenzoic, vanillic, azelaic, and malonic acids and aldehydes such as p-hydroxybenzaldehyde, vanillin, and hydroquinone were identified in the aqueous extract. On the other hand, Euphrosine-Moy et al. (1991) ozonized hydrated poplar sawdust with 45 % moisture and identified oxalic and formic acids as the major products in the aqueous extract of the treated material. Also glycolic, glycoxylic, succinic, glyceric, malonic, p-hydroxybenzoic, fumaric, and propanoic acids were identified.

Sulfur Trioxide

Yao et al. (2011) explored the use of sulfur trioxide in a process called sulfur trioxide microthermal explosion (STEX) to pretreat biomass such as rice straw. Biomass is hanged above a solution of oleum and sodium hydroxide (1 % w/v) and swirled in a test tube at 50 °C/1 atm for 7 h, followed by washing to obtain the solids (Yao et al. 2011). The internal explosion is believed to take place due to heat released from sulfur trioxide–straw reaction that causes rapid expansion of air and water from the interior of the biomass which results in enhanced structural changes and pore volume and partial removal of hemicellulose and lignin (Yao et al. 2011; Li et al. 2012a). Pretreatment of rice straw at the above-mentioned conditions resulted in the saccharification yield of 91 % (Yao et al. 2011). The efficient handling of sulfur trioxide will be a major challenge due its corrosiveness regarding this emerging pretreatment.

4.1.3.2 Acid Treatment

Concentrated acids (sulfuric acid and hydrochloric acid) have been used to treat lignocellulosic materials (Agbor et al. 2011; Bensah and Mensah 2013; Chaturvedi and Verma 2013; Kumar et al. 2009). Although they are powerful agents for cellulose hydrolysis, concentrated acids are toxic, corrosive, and hazardous and require reactors that are resistant to corrosion. To make the process economically feasible, the concentrated acid must be recovered after hydrolysis (Sivers and Zacchi 1995). Dilute acid hydrolysis has been successfully used for pretreatment of lignocellulosic materials. It has been used in case of wide range of feedstocks, including softwood, hardwood, herbaceous crops, agricultural residues, wastepaper, and municipal solid waste. It performed well on most biomass materials. Hydrochloric acid, nitric acid, phosphoric acid, and sulfuric acid have been examined for use in biomass pretreatment. Dilute sulfuric acid is commonly used as the acid of choice (Kim et al. 2000; Mosier et al. 2005b; Nguyen et al. 2000; Torget et al. 1992). It is mixed with biomass to solubilize hemicellulose thereby increasing the accessibility of the cellulose in the biomass. This mixture can be heated directly with the use of steam as in steam pretreatment, or indirectly via the vessel walls of the reactor. The dilute sulfuric acid pretreatment can significantly improve cellulose hydrolysis and can obtain high reaction rates (Esteghalian et al. 1997). At moderate temperature, direct saccharification suffered from low yields because of sugar decomposition. High temperature in dilute acid treatment is favorable for cellulose hydrolysis (McMillan 1994a, b). Recently developed dilute acid hydrolysis processes use less-severe conditions and achieve high xylan to xylose conversion yields. Achieving high xylan to xylose conversion yields is necessary to achieve favorable overall process economics because xylan accounts for up to a third of the total carbohydrate in many lignocellulosic materials (Hinman et al. 1992).

There are basically two types of dilute acid pretreatment processes:

- High temperature (temperature greater than 160 °C), continuous-flow process for low solids loading (5–10 % (weight of substrate/weight of reaction mixture) (Brennan et al. 1986; Converse et al. 1989)
- Low temperature (temperature less than 160 °C), batch process for high solids loading (10–40 %) (Cahela et al. 1983; Esteghalian et al. 1997).

Although dilute acid pretreatment can significantly improve the cellulose hydrolysis, its cost is usually higher than some physicochemical pretreatment processes such as steam explosion or AFEX. A neutralization of pH is required for the downstream enzymatic hydrolysis or fermentation processes.

Dilute acid pretreatment is conducted at temperature ranging from 140 to 215 °C. The residence time varies from a few seconds to minutes depending on the temperature of the pretreatment. An aqueous solution of substrate is heated to the desired temperature and pretreated using preheated sulfuric acid (concentrations of more than 4 wt%) in a stainless-steel reactor (Esteghalian et al. 1997; Torget et al. 1990). In another method preconditioned and physically pretreated biomass in a wire mesh is submerged in a circulating bath of dilute sulfuric acid. The bath is then heated to desired temperatures to effect pretreatment at different severities. Nguyen et al. (2000) reported a two-stage dilute acid pretreatment to maximize sugar recovery and improve biomass digestibility. A low temperature, low-acid concentration dilute acid pretreatment was used in the first stage to promote hemicellulose hydrolysis/recovery and a high-severity second stage was used to hydrolyze a portion of the remaining cellulose to glucose in the second stage (Nguyen et al. 2000). The acid in dilute acid pretreatment releases oligomers and monomeric sugars by affecting the reactivity of the biomass carbohydrate polymers polymers. Depending on the combined severity of the pretreatment, the sugars can be converted to aldehydes such as furfural and hydroxymethyl furfural. In a flowthrough dilute acid pretreatment process, the biomass is mixed or brought in contact with dilute sulfuric acid of less than 0.1 % to hydrolyze hemicellulose in biomass compared to the typical 0.6–3.0 % for dilute acid. A glucose yield of higher than 80 %, at a sulfuric acid equivalent of 0.05 wt% was reported by Mok and Antal (1992).

The addition of dilute sulfuric acid to cellulosic material has been used to commercially produce furfural which is purified by distillation and is being marketed (Zeitsch 2000). Dilute acid has been applied in the pretreatment of corn residues and short rotation woody and herbaceous crops (Torget et al. 1990, 1992). The technique is being developed by NREL for subsequent commercial application. A potential two-stage dilute acid pretreatment of softwoods chips with the objective of improving the digestibility of feedstocks with high lignin content was proposed for large-scale application (Nguyen et al. 2000).

Table 4.6 shows the advantages and disadvantages of acid treatment.

4.1.3.3 Alkali Treatment

Alkaline pretreatment is one of the major chemical pretreatment technologies. It uses various bases, including sodium hydroxide, potassium hydroxide, calcium hydroxide (lime), aqueous ammonia, ammonia hydroxide and sodium hydroxide in combination with hydrogen peroxide or others. Pretreatment with alkali cause swelling of biomass which increases the internal surface area of the biomass and decreases both the degree of polymerization and cellulose crystallinity. Alkaline pretreatment disrupts the lignin structure and breaks the linkage between lignin and the other carbohydrate fractions in lignocellulosic biomass, thus making the

Table 4.6 Advantages and disadvantages of dilute acid pretreatment

Advantages
Can achieve high reaction rates
Significantly improve hemicellulose and cellulose hydrolysis by varying the severity of the pretreatment, so the concept of combined severity can be conveniently applied
Disadvantages
Formation of degradation products and release of natural biomass fermentation inhibitors are other characteristics of acid pretreatment
Acidic prehydrolyzates must be neutralized before the sugars proceed to fermentation
Gypsum has problematic reverse solubility characteristics when neutralized with inexpensive calcium hydroxide
Disposal of neutralization salts is needed
Biomass particle size reduction is necessary. Plus, the current sulfuric acid price has increased quickly so that the economically feasibility of dilute acid pretreatment might need to be reconsidered
Dilute acid pretreatment costs more than most other physicochemical pretreatment methods, such as SP and AFEX, especially the two-stage dilute acid pretreatment
Corrosion caused by dilute acid pretreatment with sulfuric acid mandates expensive construction material
Counter cost of nitric acid negates the possibility of using it as a less corrosive replacement for sulfuric acid to reduce containment cost

carbohydrates in the heteromatrix more accessible. The reactivity of remaining polysaccharides increases as the lignin is removed. Acetyl and other uronic acid substitutions on hemicellulose that reduce the accessibility of enzymes to cellulose surface are also removed by alkali pretreatments (Chandra et al. 2007; Chang and Holtzapple 2000; Galbe and Zacchi 2007; Mosier et al. 2005b). However, most of the alkali is utilized. The effectiveness of alkaline pretreatment varies, depending on the substrate and treatment conditions. In general, alkaline pretreatment is more effective on hardwood, herbaceous crops, and agricultural residues with low lignin content than on softwood with high lignin content (Bjerre et al. 1996).

Millet et al. (1976) observed that the digestibility of sodium hydroxide-treated hardwood increased from 14 to 55 % with the decrease of lignin content from 24–55 to 20 %. However, slight effect of dilute sodium hydroxide pretreatment was found for softwoods with lignin content higher than 26 %. Kim and Holtzapple (2005) used lime to pretreat corn stover and obtained maximum lignin removal of 87.5 % at 55 °C for 4 weeks with aeration. Using lime pretreatment at ambient conditions for up to 192 h, Playne (1984) improved the enzyme digestibility of the sugarcane bagasse from 20 to 72 %. He also concluded that lime would be the chemical of choice based on the cost of chemicals. Using alkali (sodium hydroxide, calcium hydroxide and potassium hydroxide) to pretreat rice straw in 24 h at 25 °C, the authors found that sodium hydroxide (6 % chemical loading, g/g dry rice straw) was the best alkali to obtain 85 % increase of glucose yield by enzymatic hydrolysis. Aqueous ammonia is also a common alkali chemical for alkaline pretreatment.

The digestibility of sodium hydroxide-treated hardwood increased from 14 to 55 % with the decrease of lignin content from 24–55 to 20 %. However, no effect of dilute sodium hydroxide pretreatment was observed for softwoods with lignin content higher than 26 % (Millet et al. 1976). Dilute sodium hydroxide pretreatment was also effective for the hydrolysis of straws with a relatively low lignin content of 10–18 % (Bjerre et al. 1996).

Combination of irradiation and 2 % sodium hydroxide for pretreatment of corn stalk, cassava bark and peanut husk was examined by Chosdu et al. (1993). They used a combination of irradiation and 2 % sodium hydroxide for pretreatment of corn stalk, cassava bark, and peanut husk. The glucose yield of corn stalk was 20 % in untreated samples compared to 43 % after treatment with electron beam irradiation at a dose of 500 kGy and 2 % sodium hydroxide, but the glucose yields of cassava bark and peanut husk were only 3.5 and 2.5 %, respectively.

Hu et al. (2008), Hu and Wen (2008) used radio frequency-based dielectric heating during the sodium hydroxide pretreatment of switchgrass to improve its enzymatic digestibility. Switchgrass could be treated on a large-scale, at high solids content, and with a uniform temperature profile because of the unique features of radio frequency heating which are volumetric heat transfer, deep heat penetration of the samples, etc. At 20 % solids content, radio frequency-assisted alkali pretreatment (at 0.1 g of sodium hydroxide/g of biomass loading and 90 °C) resulted in a higher xylose yield than the conventional heating pretreatment. The optimal particle size and alkali loading in the radio frequency pretreatment were determined to be

0.25–0.50 mm and 0.25 g of sodium hydroxide/g of biomass, respectively. At alkali loadings of 0.20–0.25 g of sodium hydroxide/g of biomass, a heating temperature of 90 °C, and a solids content of 20 %, the glucose, xylose, and total sugar yields from the combined radio frequency pretreatment and enzymatic hydrolysis were 25.3, 21.2, and 46.5 g/100 g of biomass, respectively.

Hu et al. (2008) soaked switchgrass in sodium hydroxide solutions of different concentrations and then treated the samples by microwave or conventional heating. With alkali loadings of 0.05–0.3 g of alkali/g of biomass, microwave pretreatment resulted in higher sugar yields than conventional heating, with the highest yield (90 % of maximum potential sugars) was obtained at an alkali loading of 0.1 g/g. Table 4.7 show the advantages and disadvantages of alkali treatment.

Lime pretreatment has been reported to improve the digestibility of lignocellulosic biomass (Chang et al. 1998). It is a low-cost alkaline physicochemical pretreatment and utilizes aqueous calcium hydroxide at low temperatures and pressures as a pretreatment agent to solubilize hemicellulose and lignin (Chang et al. 1997). Calcium hydroxide is the least expensive per kilogram of hydroxide. It is possible to recover calcium from an aqueous reaction system as insoluble calcium carbonate by neutralizing it with inexpensive carbon dioxide; calcium hydroxide can subsequently be regenerated using established lime kiln technology.

Using 0.1 g calcium hydroxide/g biomass, lime pretreatment can be conducted within a wide temperature range 25–130 °C. At ambient temperatures (25 °C), the lime pretreatment could take weeks and at high temperatures (120 °C) only 2 h were required for the pretreatment of switch grass, solubilizing about 26 % xylan and 29–33 % lignin (Chang et al. 1997). The effectiveness of lime pretreatment has been attributed to opening of "acetyl valves" and "lignin valves," i.e., deacetylation and partial delignification (Chang and Holtzapple 2000). Oxidative factors come into play when oxygen is introduced at high pressures to enhance the pretreatment. Lime pretreatment of wheat straw at 85 °C for 3 h and poplar wood at 150 °C for 6 h using 14 atm oxygen resulted in better yield than without oxygen (Chang et al.

Table 4.7 Advantages and disadvantages of alkaline pretreatment

Advantages
Utilize lower temperatures and pressures than other pretreatment technologies
Can be carried out at ambient conditions, but pretreatment times are on the order of hours or days rather than minutes or seconds
Alkaline processes cause less sugar degradation compared with acid processes, and many of the caustic salts can be recovered and/or regenerated
Disadvantages
Less effective as lignin content of the biomass increases
A significant disadvantage of alkaline pretreatment is the conversion of alkali into irrecoverable salts and/or the incorporation of salts into the biomass during the pretreatment reactions so that the treatment of a large amount of salts becomes a challenging issue for alkaline pretreatment

1998, 2001). Wet oxidative treatment involves the addition of an oxidizing agent like oxygen, water, or hydrogen peroxide to biomass suspended in a liquid to improve hemicellulose and lignin removal. Lime has been also used to treat switchgrass (100 °C for 2 h) and corn stover (100 °C for 13 h) (Chang et al. 1997; Karr and Holtzapple 1998).

The process of lime pretreatment involves slurrying the lime with water, spraying it onto the biomass material, and storing material in a pile for a period of hours to weeks. The particle size of the biomass is typically 10 mm or less. Higher temperatures reduce contact time. Kim and Holtzapple (2006b) have reported that the enzymatic hydrolysis of lime-treated biomass is affected by structural features resulting from the treatment. These are the extents of acetylation, lignification, and crystallization. Lime pretreatment removes amorphous substances (e.g., lignin and hemicellulose), which increases the crystallinity index.

Chang and Holtzapple (2000) reported correlations between enzymatic digestibility and structural factors: lignin content, crystallinity, and acetyl content. They found that extensive delignification is sufficient to obtain high digestibility regardless of acetyl content and crystallinity; delignification and deacetylation remove parallel barriers to enzymatic hydrolysis; and crystallinity significantly affects initial hydrolysis rates but has less of an effect on ultimate sugar yields. Effective lignocellulose treatment process should remove all of the acetyl groups and reduce the lignin content to about 10 % in the treated biomass. Alkaline pretreatment can play a significant role in exposing the cellulose to enzyme hydrolysis. Lignin removal increases enzyme effectiveness by eliminating nonproductive adsorption sites and by increasing access to cellulose and hemicellulose.

Kim and Holtzapple (2006b) pretreated corn stover with excess calcium hydroxide (0.5 g of calcium hydroxide/g of raw biomass) in both nonoxidative and oxidative conditions at 25, 35, 45, and 55 °C. The enzymatic digestibility of lime-treated corn stover was affected by the change of structural features such as acetylation, lignification, and crystallization resulting from the treatment. Extensive delignification required oxidative treatment and additional consumption of lime (up to 0.17 g of calcium hydroxide/g of biomass). Deacetylation reached a plateau within 1 week, and there were no significant differences between nonoxidative and oxidative conditions at 55 °C; both conditions removed about 90 % of the acetyl groups in 1 week at all temperatures studied. Delignification highly depended on temperature and the presence of oxygen. Lignin and hemicellulose were selectively removed or solubilized, but cellulose was not affected by lime pretreatment at mild temperatures (25–55 °C). The degree of crystallinity increased slightly with delignification (from 43 to 60 %) because amorphous components such as lignin and hemicellulose were removed.

Lee and Fan (1982) observed that the rate of enzymatic hydrolysis depends on enzyme adsorption and the effectiveness of the adsorbed enzymes, instead of the diffusive mass transfer of enzyme. Lignin removal increases enzyme effectiveness by eliminating nonproductive adsorption sites and by increasing access to cellulose and hemicellulose.

Kong et al. (1992) reported that alkalis remove acetyl groups from hemicellulose (mainly xylan), thereby reducing the steric hindrance of hydrolytic enzymes and greatly enhancing carbohydrate digestibility. They concluded that the sugar yield in enzymatic hydrolysis is directly associated with acetyl group content.

Karr and Holtzapple (2000) showed that pretreatment with slake lime (calcium hydroxide) increased the enzymatic hydrolysis of corn stover by a factor of 9 compared to that of untreated corn stover. The optimal pretreatment conditions were determined to be a lime loading 0.075 g of calcium hydroxide/g of dry biomass, a water loading of 5 g of water/g of dry biomass, and heating for 4 h at 120 °C. It was suggested that pretreatment with lime can lead to corn stover polysaccharide conversions approaching 100 %. Dilute sodium hydroxide treatment of lignocellulosic materials has been found to cause swelling, leading to an increase in internal surface area, a decrease in the degree of polymerization, separation of structural linkages between lignin and carbohydrates, a decrease in crystallinity and disruption of the lignin structure (Fan et al. 1987). The digestibility of sodium hydroxide-treated hardwood was reported to increase from 14 to 55 % with a decrease of lignin content from 24–55 to 20 %. However, no effect of dilute sodium hydroxide pretreatment was observed for softwoods with lignin content greater than 26 %. Dilute sodium hydroxide pretreatment was also found to be effective for the hydrolysis of straws with relatively low lignin contents of 10–18 % (Bjerre et al. 1996).

Lime pretreatment has been used in lab scale. Conversely, oxidative effect of adding water, air, or oxygen could be used in an industrial pretreatment process to improve biomass digestibility by percolating oxygen through a pile of biomass.

Table 4.8 shows the advantages and disadvantages of lime treatment.

4.1.3.4 Organosolv Process

Organosolv process has been extensively utilized for extraction of high-quality lignin, which is a value-added product, and once the lignin is removed from the biomass, the cellulose fibers become accessible to enzymes for hydrolysis and absorption of cellulolytic enzymes to lignin is minimized which leads to higher conversion of biomass (Agbor et al. 2011).

The organosolv pretreatment process involves the addition of an aqueous organic solvent mixture to the biomass under specific temperatures and pressures with/without a catalyst (Sun and Cheng 2002; Alriols et al. 2009; Chum et al. 1985). The catalysts used are acid, a base, or a salt and the solvents which are most commonly used are ethanol, methanol, ethylene glycol and acetone (Ichwan and Son 2011). Temperatures used in the process can be as high as 200 °C, but lower temperatures can also be used depending on the type of biomass and the use of a catalyst. The process produces three main fractions—a high purity lignin, a relatively pure cellulose fraction, and a hemicellulosic syrup containing C5 and C6 sugars. The pretreated solid residues are separated by filtration and washed with distilled water in order to remove solvents and degradation products which may

Table 4.8 Advantages and disadvantages of lime pretreatment

Advantages
Low reagent cost, safety, and ease of recovery as calcium carbonate compared with sodium, potassium, and ammonium hydroxide even though sodium hydroxide has receive the most attention
Lime pretreatment performed at temperatures below 100 °C avoids the huge energy demands required to maintain high thermal steady-state conditions during pretreatment and use of pressured vessels
Huge piles of biomass could simply be pretreated without the need for a special vessel or using a simple design for pilot plants
Recalcitrant biomass is rendered digestible as result of the combined action of the alkali and oxygen. About 80 % of the lignin in high-lignin lignocellulose material such as hardwoods (e.g. poplar wood) is removed during oxidative lime pretreatment
Disadvantages
Lime pretreatment is not very effective for removing lignin in high-lignin biomass such as softwood
Action of lime is slower than ammonia and most often the oxidizing agent is not very selective as a result losses in hemicellulose and cellulose may occur
Use of huge volumes of water in the washing step (5–10 g water/g biomass) and the potential need for neutralization makes downstream processing difficult and also increase the cost of scaling-up lime pretreatment
The oxidation of lignin to other soluble aromatic compounds is a risk due to the possibility of the formation of inhibitors

inhibit downstream process such as enzymatic hydrolysis and fermentation. Pretreatment conditions lead to simultaneous hydrolysis and lignin removal. This occurs via the disruption of internal bonds in lignin, lignin-hemicellulose bonds, and glycosidic bonds in hemicelluloses and to a smaller extent in cellulose (Conde-Mejía et al. 2012; Chum et al. 1985). Other changes include the formation of droplets of lignin on the surface of pretreated biomass, a situation that inhibits hydrolysis by adsorption on the surface of the cellulose (Koo et al. 2012). The presence of acids as a catalyst causes acid-catalyzed degradation of the monosaccharides into furfural and 5-hydroxymethyl furfural followed by condensation reactions between lignin and these reactive aldehydes (Chum et al. 1985). After removal of lignin, the biomass (rich in cellulose) is used for enzymatic hydrolysis (Zhao et al. 2009).

Process variables such as temperature, reaction time, solvent concentration, and acid dose affect the physical characteristics (crystallinity, fiber length and degree of polymerization of cellulose) of the pretreated substrates. In most cases, high temperatures and acid concentrations and also long reaction times cause considerable degradation of sugar into fermentation inhibitors. Sulfuric acid has been used extensively as a catalyst due to its strong reactivity but, it is corrosive, possesses inhibitory characteristics and toxic (Park et al. 2010).

Park et al. (2010) evaluated the effectiveness of sulfuric acid, sodium hydroxide and magnesium sulfate catalysts on pine and found the acid as the most effective in terms of the ethanol yield. But an increase in the concentration of the base from 1 to 2 % had a positive effect on digestibility. The use of carbon dioxide as a catalyst did not improve process yields on pretreatment of willow wood (Huijgen et al. 2008). Organic solvents/acids that have been used as catalysts include oxalic, formic, acetylsalicylic, salicylic acid, ethanol, methanol, acetone, ethylene glycol, triethylene glycol, and tetrahydrofurfuryl alcohol (Sun and Cheng 2002; Haverty et al. 2012; Mesa et al. 2011; Zhao and Liu 2012).

The Battelle organosolv method involves the use of a ternary mixture of phenol, water, and hydrochloric acid to fractionate the biomass at about 100 °C and at 1 atm (Villaverde et al. 2010). The acid depolymerizes lignin and hydrolyses the hemicellulose fraction and the lignin dissolves in the organic phase (phenol), while the monosaccharides are accumulated in the aqueous phase upon cooling of the fractionated biomass. Similarly, the formic acid organosolv process (formasolv) involves the application of formic acid, water, and hydrochloric acid to depolymerize, oxidize, and dissolve lignin, hemicellulose, and extractives in the biomass, and the precipitation of the lignin is obtained by the addition of water (Villaverde et al. 2010). Formic acid has a good lignin solvency and the process can be conducted under low temperatures and at atmospheric pressure (Zhao and Liu 2012). Formic acid, however, causes formylation of the pretreated substrates which could reduce cellulose digestibility. Pretreated substrates can be deformylated in alkaline solution as was observed on bagasse at 120 °C (Zhao and Liu 2012).

Kupiainen et al. (2012) used formic acid (5–10 % w/w; no catalysts) on delignified wheat straw pulp at 180–220 °C, yielding a maximum glucose yield of 40 %. Organosolv pretreatment with acetic acid (acetosolv) produces higher yields than formic since less material is dissolved for a given time. Higher cellulose viscosity in smaller time periods is obtained in acetosolv process (Villaverde et al. 2010).

The use of ethanol in organosolv pretreatment enables the recovery of high value products which include cellulose, sulfur- and chlorine-free lignin, enriched hemicelluloses, and extractives. Via the use of solvents such as ionic liquids, further purification may be obtained (Prado et al. 2012). The ethanosolv process is usually operated under higher pressures and temperatures unlike formasolv process. In addition, reprecipitation of lignin takes place due to lower lignin solubility (Zhao and Liu 2012). The reduced toxicity of ethanol compared to solvents such as methanol to the downstream fermentation process and the fact that ethanol is the final product are additional benefits (Chum et al. 1985; Kim et al. 2011). Generally, lower ethanol/water ratios favor hemicellulose hydrolysis and the enzymatic degradability of pretreated biomass since ethanol inhibits the performance of hydrolytic enzymes (Huijgen et al. 2008). Ethanosolv process has been explored for the development of proprietary technologies such as the Alcell process and the Lignol process. Alcell process is a sustainable alternative to kraft pulping (Pye and Lora 1991) and the Lignol process—a biorefinery platform that uses aqueous ethanol (50 % w/w) for pretreating lignocellulosic biomass at 200 °C and 400 psi to separate the different components in biomass (Arato et al. 2005; Pan et al. 2005). High sugar yields and product recovery have been observed in case of ethanosolv pretreatment of various materials including hybrid poplar (Pan et al. 2006a) and Japanese cypress (Hideno et al. 2013). A main advantage is the potential to recover much of the ethanol (Koo et al. 2012; Alriols et al. 2010) and water (Alriols et al. 2009). This reduces the operating cost. Ethanosolv coproducts such as hemicellulose syrup and lignin can serve as feedstocks for the production of high-value products. Furthermore, ethanosolv lignins whose molecular weight and functional groups depend on process conditions are known to possess antioxidant properties (Pan et al. 2006b). In several situations, presoaking materials, for example, in acidic medium (Kim et al. 2012) or in bioslurry (Brosse et al. 2010), positively affect the process in terms of sugar yields and lignin removal, among seven others. Other variations involve the combined use of acid and basic catalysts, microwave-assisted organosolv, and the avoidance of catalysts (Mesa et al. 2011; Li et al. 2012b; Wang et al. 2012). In another variation, the biomass is treated with the inclusion of ferric sulfate and sodium hydroxide to the biomass/liquor (formic acid and hydrogen peroxide) mixture. Formic acid reacts with hydrogen peroxide to form peroxyformic acid and its application in organosolv pretreatment (Milox) of biomass produced good results on hardwoods and also softwoods (Villaverde et al. 2010).

Organosolv process was studied by Mesa et al. (2011) for extraction of sugarcane bagasse. 30 % (v/v) ethanol for 60 min at 195 °C produced 29.1 % of reducing sugars under optimized conditions. The scale-up of the process, by performing the acid pretreatment in a 10-L semi-pilot reactor fed with direct steam, was successfully conducted. Pan and Saddler (2013) used extracted lignin obtained from ethanol organosolv and used it as a replacement of petroleum—based polyol for the production of Rigid Polyurethane Foams. The results showed that the prepared foams contained 12–36 % (w/w) extracted lignin. The compressive strength, density, and cellular structure of the prepared foams were studied and compared. The researchers observed that the lignin-containing foams had comparable structure and strength with polyol containing foams.

Geng et al. (2012) studied organosolv pretreatment of horticultural waste for bioethanol production. A modified method using ethanol under mild conditions followed by hydrogen peroxide post-treatment was studied for horticultural waste. Enzymatic hydrolysis of the organosolv pretreated horticultural waste with 17.5 % solid content, enzyme loading of 20 FPU/g horticultural waste of filter paper cellulase, and 80 CBU/g horticultural waste of beta-glycosidase resulted in a horticultural waste hydrolysate containing 15.4 % reducing sugar after 72 h. Fermentation of the above hydrolysate medium produced 11.69 g/L ethanol at 8 h using Saccharomyces cerevisiae.

Sun and Chen (2008) studied pretreatment of wheat straw by glycerol-based autocatalytic organosolv pretreatment. Under optimized conditions (liquid–solid ratio of 20 g/g at 220 °C for 3 h), 70 % hemicelluloses and 65 % lignin were removed from the biomass and also, 98 % cellulose retention was obtained. The pretreated fibers showed 90 % of the enzymatic hydrolysis yield after 48 h of incubation.

Hideno et al. (2013) have shown that application of alcohol-based organosolv treatment in combination with ball milling to Japanese cypress (Chamaecyparis obtusa) significantly improved the enzymatic digestibility and decreased the required severity of organosolv treatment. Moreover, alcohol-based organosolv treatment increased the efficiency and reduced the time required for ball milling despite small quantity of removed lignin. It was found that the combination of alcohol-based organosolv treatment in mild condition and short-time ball milling had a synergistic effect on the enzymatic digestibility of Japanese cypress.

Ichwan and Son (2011) studied organosolv extraction of oil palm pulp using various solvents such as ethanol–water, ethylene glycol–water, and acetic acid–water mixture to extract cellulose. The yield of organosolv pulping with ethylene glycol–water, ethanol–water, and acetic acid—water mixture was 50.1; 48.1, and 41.7 %, respectively, whereas the kappa number was 74.7; 72.7; and 67. Fourier Transform Infrared Spectroscopy spectra showed that the degradation of cellulose pulp occurred during acid pulping. The X-ray diffraction measurement showed that the ethanol–water mixture pulping resulted in higher crystallinity of pulp (68.67), followed by ethylene glycol–water mixture as (58.14), acid pulping (54.21).

Panagiotopoulos et al. (2012) used steam followed by organosolv treatment to separate hemicellulose, lignin and cellulose components from poplar wood chips. Lignin extraction was enhanced as more than 66 % of lignin was extracted. More than 98 % of the original cellulose was recovered after the two-stage pretreatment and 88 % of the cellulose could be hydrolyzed to glucose at enzyme loading of 5 FPU/g of cellulose after 72 h.

Table 4.9 shows the advantages and disadvantages of Organosolv process (Agbor et al. 2011).

4.1.3.5 Sulfite Pretreatment

Sulfite application is emerging as a promising biomass pretreatment method due to positive results obtained from several materials (Bensah and Mensah 2013). University of Wisconsin, Madison researchers have developed an improved pre-treatment process for conversion of biomass. This process, known as Sulfite Pretreatment to Overcome Recalcitrance of Lignocellulose (SPORL), reduces the energy consumption needed for size-reduction processes, required before enzymatic hydrolysis, by more than tenfold. The new method can use several aqueous sulfite or bisulfite solutions over a wide range of pH values and temperatures to weaken the chemical structure of the plant material. The improved SPORL method is flexible and can be integrated easily into current pretreatment systems. It is particularly suitable for woody biomass, softwoods such as pines and other conifers and hardwoods such as poplar, willow and eucalyptus. The pH of the pretreatment liquor can be adjusted by reagent, making SPORL easily incorporated into current dilute acid approaches to improve the efficiency of the pretreatment. Depending on

Table 4.9 Advantages and disadvantages of organosolv process

Advantages
Selective pretreatment method
Organosolv lignin is sulfur-free with high purity and low molecular weight. It can be used as fuel to power pretreatment plant or further purified to obtain high-quality lignin which is used a substitute for polymeric materials used for the manufacture of bioplastics
Very effective for the pretreatment of high-lignin lignocellulose materials, such as soft woods
This process can be combined with other pretreatment method to obtain a clean/effective biomass fractionation process for more recalcitrant biomass as means to improve pretreatment yield
Organic solvents are easily recovered by distillation and recycled for reuse as a means to reduce chemical consumption and the cost of the process
This process does not require significant size reduction of feedstock to achieve satisfactory cellulose conversion making the process less energy intensive especially with the pretreatment of woody biomass
Disadvantages
Cost of chemicals and sometimes catalyst makes this process more expensive than other leading pretreatment processes
Side reactions such as acid-catalyzed degradation of monosaccharides into furfural and 5-HMF that are inhibitory to fermentation microorganisms, have been associated with this process
Using volatile organic liquid at high temperature necessitates using of containment vessels thus; no digester leaks can be tolerated due to inherent fire, explosion hazards, environmental and health and safety concerns

the stock material, mechanical size reduction steps such as disk or hammer milling can be implemented directly before or after SPORL. In addition, the final enzymatic hydrolysis can be coupled directly after the pretreatment with or without washing the material or adding a surfactant to aid in the process. The pretreatment also can be used with steam explosion, using bisulfite as a catalyst. The hydrolyzed biomass can be separated and the sugars fermented or catalytically converted into fuels after pretreatment and the sulfonated lignin byproducts can be sold and other wastes burned to produce energy for the process.

Table 4.10 shows the applications of SPORL process. The SPORL method is a superior method of biomass pretreatment due to its versatility, efficiency, and simplicity (Table 4.11). It shows excellent scalability to commercial production. Through decreased size-reduction energy requirements and maximized enzymatic cellulose conversion in a short period of time, this method will increase the energy efficiency of ethanol fermentation and catalytic fuel production processes. This

Table 4.10 Applications of SPORL process

Applications of SPORL process
Production of ethanol, biofuels, biochemicals, or other bioproducts
Pretreatment of biomaterials with strong physical integrity
Recycling agricultural and forestry product wastes

Table 4.11 Key benefits of SPORL process

Key benefits of SPORL process
Reduces energy consumption of size-reduction processes by more than tenfold
Couples directly with enzymatic hydrolysis processes without further treatment
Approximately 90 % conversion of the cellulose fraction to glucose
Maximizes hemicellulose and lignin removal and recovery
Minimizes cellulose lost by degradation
Minimizes cost of chemicals and reagents
Able to process a variety of lignocellulosic biomaterials
Able to retrofit existing pulping mills or equipment to produce biofuels
Excellent scalability for building new cellulosic ethanol plants
Sulfonated lignin byproducts can be sold as dispersants and extenders

increase in efficiency will allow biofuels and other bioproducts to become economically competitive with petroleum-derived fuels and products.

In this method, milled biomass (<6 mm) is mixed in (1–10 % w/w) sulfite (sodium sulfite, sodium bisulfate) solution in acidic, basic, or neutral environments at selected temperatures (80–200 °C) and reaction times (30–180 min). Using filtration, pretreated solids are separated from the spent liquor. The solids are washed with distilled water and dried before undergoing saccharification. The process degrades and sulfonates lignin partially and improves glucose yields due to the formation of sulfonic and weak acid groups. This improves the hydrophilicity of pretreated substrates (Bu et al. 2012). Sulfonation is increased in the presence of volatile organic solvents. These solvents reduce the surface tension and allow effective penetration of the solution in the biomass. In addition to this, the lignin is hydrolyzed and dissolved in the organic phase and can be easily recovered in pure forms (Bu et al. 2012). Lignin removal can be further improved by the presence of sulfomethyl groups produced from the combined action of sulfite and formaldehyde on lignin, resulting in high sugar yields (Jin et al. 2013). The process was investigated with the dried wood chips and the chops of corn stalks impregnated with a bisulfite solution, ranging from a pH of 1–10, at a temperature of 180 °C for 30 min, immediately followed by mechanical size-reduction and enzymatic hydrolysis. It was found that almost all hemicellulose and 20 % of the lignin were removed at an optimal pH of 2 and temperature of 180 °C. About 90 % of the recovered cellulose could be used in hydrolysis when feedstock was treated with a sulfite pretreatment solution at a pH of 2 for 48 h at a normal cellulase dose level of about 5–20 FPU/g cellulose. Use of polyethylene glycol during hydrolysis gave cellulose conversions of approximately 90 % in 12 h and increased to 95 % in 48 h.

The SPORL process has been effectively used to pretreat lodgepole pine for subsequent conversion to ethanol at a yield of 276 L/t wood and at a net energy output of 4.55 GJ/t wood (Zhu et al. 2010b). The conditions used were—liquor/wood ratio = 3:1 v/w; temperature −180 °C and RT—25 min). SPORL pretreatment of switchgrass was found to be superior to dilute acid and alkali (Zhang et al. 2013b; Shuai et al. 2010) in terms of the digestibility of the pretreated substrates. Similarly, higher sugar yields and lower inhibitor concentration were found with SPORL-pretreated agave stalk relative to dilute acid and sodium hydroxide (Yang and Pan 2012). The SPORL process was also found superior to the organosolv and steam explosion pretreatments based on total sugar recovery and energy consumption (Zhu and Pan 2010).

Pretreatment of corn stover with alkaline sodium sulfite at 140 °C resulted in 92 % lignin removal and 78.2 % total sugar yield (0.48 g/g raw biomass) after enzymatic hydrolysis, which was higher than four other alkali-based methods under similar conditions (Li et al. 2012c). About 79.3 % of total glucan was converted to glucose and cellobiose during corn cob pretreatment at 156 °C, 1.4 h, 7.1 % charge, and solids loading of 1:7.6 w/w; and subsequent SSF gave 72.2 % theoretical ethanol yield (Cheng et al. 2011).

In another study by Bu et al. (2012), pretreatment of corn cob residues produced the highest glucose yields (81.2 %) in the presence of ethanol compared to acidic, basic, and neutral conditions. At the pilot level, sulfite-based methods have been studied and showed good results.

Liu and Zhu (2010) reported that the negative effects of soluble inhibitors and lignosulfonate on enzymatic hydrolyses could be counteracted by adding metal salts to the pretreated contents, making it possible to avoid the costly washing process.

Table 4.12 shows advantages and disadvantages of sulfite pretreatment (Bensah and Mensah 2013).

Table 4.12 Advantages and disadvantages of sulfite pretreatment

Advantages
High sugar yields
Lignin removal
Recovery of biomass components in less chemically transformed forms
Disadvantages
Sugar degradation at severe conditions
Large volumes of process water used in post pretreatment washing
High costs of recovering pretreatment chemicals

4.1.4 Cellulose Solvent-Based Lignocellulose Pretreatments

Cellulose solvent-based lignocellulose pretreatments are gaining attention because they can break the recalcitrant structure of biomass by increasing cellulose accessibility in a more effective manner than traditional biomass pretreatments (Sathitsuksanoh et al. 2012a). The hydrolysis rate and digestibility of pretreated biomass are increased as a result of this and enzyme use is decreased (Sathitsuksanoh et al. 2009, 2010, 2012a, b). Also, cellulose solvent-based pretreatments may be regarded as a biomass-independent pretreatment. A few cellulose solvent-based strategies are being developed such as concentrated phosphoric acid [85 % (w/w)], ionic liquids, NMMO, NaOH/urea, and DMAc/LiCl.

4.1.4.1 Cellulose Solvent- and Organic Solvent-Based Lignocellulose Fractionation

Cellulose solvent- and organic solvent-based lignocellulose fractionation (COSLIF) were developed to fractionate lignocellulose using a combination of concentrated phosphoric acid as a cellulose solvent and an organic solvent such as acetone or ethanol under modest reaction conditions (Ladisch et al. 1978). The ideas of COSLIF are following (Zhang et al. 2006, 2007a, b, c; Zhang and Lynd 2006):

- Removal of partial lignin and hemicellulose—eliminating the major obstacles to hydrolysis and allowing cellulase to access the substrate more efficiently
- Decrystallization of cellulose fibers—providing better cellulose accessibility to cellulase
- Modest reaction conditions—a reduction in sugar degradation, reduced formation of inhibitors, reduced utility consumption, and capital investment.

Concentrated phosphoric acid can completely dissolve cellulose fibers. This results in effective disruption of highly ordered hydrogen bonding network of crystalline cellulose (Sathitsuksanoh et al. 2011; Moxley et al. 2008) and drastic increases in CAC (Rollin et al. 2011; Zhu et al. 2009). COSLIF has been demonstrated to efficiently pretreat a variety of feedstocks, such as bamboo, bermudagrass, common reed, corn stover, gamagrass, giant reed, elephant grass, sugarcane, hemp hurd, Miscanthus, poplar, and switchgrass (Sathitsuksanoh et al. 2009, 2010, 2012a, b; Zhu et al. 2009; Zhang et al. 2007a, b, c; Moxley et al. 2008; Conte et al. 2009; Ge et al. 2012).

Different species of untreated biomass show a large variation in their glucan digestibilities at 15 filter paper units (FPUs) of cellulase per gram of glucan, showing their different recalcitrant degrees. However, all of the COSLIF-pretreated biomass have similar high digestibilities (>87 %) after 72 h at an enzyme loading of 5 FPUs of cellulase per gram of glucan. Therefore, COSLIF treatment could be regarded as a feedstock-independent pretreatment. The cost of fungal cellulose is high approximately 100 US cents per gallon of cellulosic ethanol. Therefore,

3–5-fold reduction in cellulase use means up to 80 cents saving per gallon of ethanol produced (Sathitsuksanoh et al. 2010). Optafuel in Southern Virginia (United States) is studying COSLIF technology in a pilot plant.

Sathitsuksanoh et al. (2012a) studied the correlation between cellulose accessibility to cellulose (CAC) values of numerous feedstocks before and after pretreatment and enzymatic glucan digestibility. Untreated biomass feedstocks with different carbohydrate and lignin contents were found to have low CAC values, resulting in low enzymatic glucan digestibility (lower than 20 %). An exception is bagasse possibly because it was prepared through leaching, drying, followed by milling that may disrupt biomass fiber more efficiently than other untreated feedstocks through simple particle size reduction. Energy-intensive milling is a very efficient biomass pretreatment for increasing substrate accessibility but it is very expensive. After pretreatments, such as dilute acid, SAA, and lime, pretreated biomass samples show improved CAC values, accompanied by higher glucan digestibility. This correlation between CAC and digestibility suggests that increasing substrate accessibility for most pretreatments is important for obtaining high enzymatic glucan digestibility. When CAC values were higher than a critical value of 8 m^2 g^{-1} biomass, very high glucan digestibilities were obtained. In these cases, digestibilities were independent of CAC values. This suggested that further improvement of CAC higher than the critical value was not important. Although COSLIF very efficiently overcomes lignocellulose recalcitrance, a large volume of cellulose solvents and organic solvents are needed so that process modification and optimization must be conducted in order to make the whole process economically attractive.

4.1.4.2 Ionic Liquids

Novel pretreatment methods using ionic liquids (ILs) are creating new opportunities in the area of biomass conversion (Socha et al. 2014; Sathitsuksanoh et al. 2012a; Wu et al. 2004, 2011). ILs are salts that exist as liquids and are composed entirely of paired ions and have tuneable properties. These are being extensively studied as "green solvents" for several industrial and research applications, particularly in the area of catalysis, chemical synthesis, and separation of cellulose-based biomass (Pu et al. 2007). They are able to dissolve large amounts of cellulose at extremely mild conditions. The possibility of recovering almost 100 % of the used ILs to their initial purity makes them attractive (Heinze et al. 2005). This technology was used for direct dissolution of cellulose in the commercial Lyocell process for modern industrial fiber-making (Fink et al. 2001).

ILs convert carbohydrates in lignocellulosic materials into fermentable sugars via two main pathways (Wang et al. 2011b). One is the pretreatment of the biomass to improve its efficiency of enzymatic hydrolysis, and the other deals with the transformation of the hydrolysis process from a heterogeneous to a homogeneous reaction system by dissolution in the solvent.

ILs are recognized to facilitate more green applications in reactions and separations due to their special beneficial properties, such as high thermal stability and negligible vapor pressure. Their very low vapor pressure reduces the risk of exposure which is a clear advantage over the use of the traditional volatile solvents. ILs are increasingly being used to dissolve various lignocellulosic biomass as shown in Table 4.13.

Table 4.14 shows properties of ionic liquids. As cellulose solvents, ILs possess several advantages over regular volatile organic solvents. They show low hydrophobicity, low toxicity, broad selection of anion and cation combinations, better electrochemical stability, thermal stability, low viscosity, high reaction rates, low volatility with potentially reduced environmental impact, nonflammable property,

Table 4.13 Use of ionic liquids for dissolution of lignocellulosic biomass

Poplar Emim-Ac
Pine Amim-Cl
Eucalyptus Emim-Ac
Spruce Amim-Cl
Bagasse Emim-Ac
Switchgrass Emim-Ac
Bamboo Emim-Gly
Wheat straw Amim-Cl, Bmim-Ac
Water hyacinth Bmim-Ac
Rice husk Bmim-Cl, Emim-Ac
Rice straw Ch-Aa
Kenaf powder Ch-Ac
Cassava pulp Emim-Ac, Dmim-SO4, Emim-DePO4

Dmim-SO4: 1,3-dimethylimidazolium methyl sulfate
Emim-DePO4: 1-ethyl-3-methylimidazolium diethyl phosphate
Ch-Aa: Cholinium amino acids
Ch-Ac: Cholinium acetate
Based on Li et al. (2010a, b), Guragain et al. (2011), Muhammad et al. (2011), Papa et al. (2012), Weerachanchai et al. (2012), Perez-Pimienta et al. (2013), Hou et al. (2012), Yuan et al. (2012), Qiu et al. (2012), Ninomiya et al. (2013), Bensah and Mensah (2013)

Table 4.14 Properties of ionic liquids (ILs)

Salts consisting of large cations (mostly organic) and small anions (mostly inorganic), with a low degree of cationic symmetry
Melting point below 100 °C
Nonflammable
Liquid at room temperature
Improve antielectrostatic and fire-proof properties of wood
Low volatility and high thermal stability up to temperatures of about 300 °C
High electrical conductivity, high solvating properties, and wide electrical window

Based on Yang and Wyman (2008), Dadi et al. (2007), Mora-Pale et al. (2011), Vancov et al. (2012)

low melting points, high polarities, negligible vapor pressure consisting entirely of ions (anions and cations) (Wasserscheid and Keim 2000; Zavrel et al. 2009). Ionic liquids exist in two main forms—simple salts consisting single cations and anions, and those where equilibrium is involved (Menon and Rao 2012). The most common forms contain the imidazolium cation which can pair with anions such as acetate, sulfate, nitrate, chloride, bromide, methanoate, and triflate. ILs could be designed and developed to pretreat specific biomass under optimal conditions by combining anions and cations which can result in an estimated formulation of 10^9 ILs (Wu et al. 2011). The properties of ILs could be changed by varying the length and branching of the alkyl groups that are integrated into the cation (Holm and Lassi 2011). The ILs 1-ethyl-3-methylimidazolium glycinate (Emim-Gly) and 1-allyl-3-methylimidazolium chloride (Amim-Cl) were synthesized from various compounds for the dissolution of bamboo and wood (Muhammad et al. 2011; Zhang et al. 2013a, b), respectively. Not all characteristics of ionic liquids are favorable as solvents in pretreatment. Chloride-based ILs such as 1-butyl-3-methylimidazolium chloride (Bmim-Cl) are corrosive, hygroscopic, and toxic while others such as Amim-Cl are viscous with reactive side chains (Xie et al. 2012). Also, ILs with long akyl chains have the tendency to obstruct nonpolar active sites of enzymes because of their hydrophobic nature (Ventura et al. 2012). Others have favorable properties and have been under investigation as promising solvents. Mora-Pale et al. (2011) have reported that phosphate-based solvents show higher thermal stability and lower toxicity and viscosity than chloride-based ones. Positive outcomes have been obtained with the use of 1-ethyl-3-methylimidazolium acetate (Emim-Ac) since it is favorable to in situ enzymatic saccharification due to its enzymatic activity and biocompatibility (Li et al. 2011).

ILs are being screened for their potential to improve the digestibility of lignocellulosic biomass (Zavrel et al. 2009).

Imidazolium-based ILs effectively dissolve biomass and represent an important platform for biomass pretreatment. Although imidazolium cations are efficient, these are expensive and therefore limited in their large-scale industrial applications.

The dissolution mechanism of cellulose in ILs involves the hydrogen and oxygen atoms of cellulose hydroxyl groups in the formation of electron donor–electron acceptor complexes which interact with the ILs (Feng and Chen 2008). These ionic liquids compete with lignocellulosic components for hydrogen bonding, thus disrupting its three dimensions network (Moultrop et al. 2005). Upon interaction of the cellulose-OH and ILs, the hydrogen bonds are broken. This results in opening of the hydrogen bonds between molecular chains of the cellulose (Feng and Chen 2008). The interaction ultimately results in the dissolution of cellulose.

It is found that ILs having imidazolium or pyridinium cations paired with Cl^-, $CF_3SO_3^-$, $CF_3CO_2^-$, $CH_3CO_2^-$, $HCOO^-$, $R_2PO_4^-$ anions are able to dissolve cellulose fibers through strong basicity of hydrogen bond. The dissolution of lignocellulose in ILs disrupts the primary bonds among cellulose, hemicellulose and lignin and provide more substrate accessibility to hydrolytic enzymes. According to Li et al. (2011), with suitable choice of anti-solvents which are water, acetone, and alcohol, up to 80 % lignin and hemicellulose can be fractionated.

By the use of antisolvents, solubilized cellulose can be recovered by rapid precipitation. The recovery cellulose was found to have the same polydispersity and the degree of polymerization as the initial cellulose, but significantly different macro- and microstructure, particularly the reduced degree of crystallinity (Zhu 2008). The previously used ILs include N-methylmorpholine-N-oxide monohydrate (NMMO) (Kuo and Lee 2009a), 1-allyl-3-methylimidazolium chloride (AMIMCl) (Dadi et al. 2007; Wu et al. 2004; Zhang et al. 2005), 1-n-butyl-3-methylimidazolium chloride (BMIMCl) (Dadi et al. 2006, 2007; Liu and Chen 2006; Swatloski et al. 2002), 3-methyl-N-bytylpyridinium chloride (MBPCl), and benzyldimethyl (tetradecyl) ammonium chloride (BDTACl) (Heinze et al. 2005).

Using 1-butyl-3-methylimidazolium chloride (BMIMCl) for pretreatment, Dadi et al. (2006) reported that the initial enzymatic hydrolysis rate and yield of pretreated Avicel-PH-101 were increased by 50- and 2-fold compared with untreated Avicel. Liu and Chen (2006) obtained significant improvement of enzymatic hydrolysis yield using BMIMCl to treat raw and steam-exploded wheat straw. They found BMIMCl modified the structure of wheat straw by decreasing the degree of polymerization and crystallinity and partially solubilizing cellulose and hemicellulose. Kuo and Lee (2009a) also observed that the 1, 3-N-methylmorpholine-N-oxide (NMMO) pretreated sugarcane bagasse has about 2 times higher enzymatic hydrolysis yield as compared to untreated bagasse.

Most ILs used in biomass fractionation are imidazonium salts. Studies have shown that 1-allyl-3-methylimidazonium chloride (AMIMCl) and 1-butyl-3-methylimidazonium chloride (BMIMCl) can be used effectively as a nonderivatizing solvent for the dissolution of cellulose at temperatures below 100 °C (Zhang and Lynd 2006; Zhu et al. 2006). The cellulose fraction can be recovered by the addition of water, ethanol or acetone. The solvent can be recovered and reused by using various techniques such as reverse osmosis, pervaporation, salting out, and ionic exchange.

Ionic liquids are able to dissolve a variety of biomass with different hardness and are introduced as selective solvents of lignin and cellulose (Cao et al. 2010). In this process, biomass is solubilized in the solvent at 90–130 °C and ambient pressure, followed by the addition of water to precipitate the biomass. The process is completed by washing the precipitate. The structure of lignin and hemicellulose remain unchanged after treatment with ionic liquids. This allows the selective extraction of unaltered lignin because lignin is highly soluble in solvents whereas the solubility of cellulose is low (Zhang et al. 2012). This can result in the separation of lignin and the enhancement in cellulose accessibility under ambient temperature and pressure without alkaline or acidic reagents and the formation of inhibitors. These solvents are expensive. But, the cost of their recovery is not high because of the low vapor pressure of the solvents. Nevertheless, cellulase enzyme is irreversibly inactivated in ionic liquid solvents (Zhi et al. 2012), which reduces the biomass conversion efficiency and increases the overall cost. This result shows the need to develop solvents in which cellulase and microorganisms are active.

Application of ionic liquids has opened new ways for the effective utilization of lignocellulosic materials in areas such as biomass pretreatment and fractionation. However, there are still many challenges in putting these potential applications into practical use. These are listed below:

- High cost of ILs
- Action mode on hemicellulose and/or lignin contents of lignocellulosic materials
- Regeneration requirement
- Lack of toxicological data and knowledge about basic physicochemical characteristics
- Generation of inhibitors

Further research and financial support are required to address these challenges.

4.1.4.3 Aqueous *N*-Methylmorpholine-*N*-Oxide

Aqueous *N*-Methylmorpholine-*N*-Oxide (NMMO) has attracted interest for use as a pretreatment solvent. It is a well-known industrial solvent used in the Lyocell process for the production of fibers. Cellulose dissolves without derivatization in NMMO/H2O system and the hydrogen bonds in cellulose are broken down in favor of new bonds between cellulose and solvent molecules (Zhao et al. 2007). This leads to swelling and increased porosity (Li 2012a, b, c), and also reduced degree of polymerization and crystallinity which improves enzymatic saccharification. Addition of boiled distilled water to pretreatment slurry containing dissolved biomass causes cellulose I to precipitate into cellulose II which is more reactive. Regenerated solids are filtered and washed with warm/boiling water until the filtrate is clear. Like ionic liquids, NMMO dissolves biomass with no/less chemical modification at low/moderate temperatures (80–130 °C). Additional favorable characteristics of NMMO pretreatment include high sugar yields, high solvent recovery, formation of low degradation products, and no harmful effect on the environment. Further, cellulase activity is not negatively affected by low concentrations (15–20 % w/w) of NMMO, indicating the potential of application in continuous processes (Li et al. 2012d).

Aqueous NMMO has already being used on several types of biomass as a sole pretreatment method or in combination with others. Poornejad et al. (2013) compared the effectiveness of NMMO and the ionic liquid (Bmim-Ac) treatment on rice straw at 120 °C/5 h. Glucan conversion was complete with the ionic acid, whereas 96 % conversion was obtained with the use of NMMO. Upon SSF, the yield of ethanol was higher with NMMO (93.3 %) than Bmim-Ac (79.7 %). Shafiei et al. (2010) observed total conversion of cellulose to ethanol at ethanol yields of up to 85.4 and 89 % for pretreatment (130 °C/3 h) of oak and spruce, respectively. High saccharification yields (>90 %) were observed for ultrasound-assisted NMMO treatment of sugarcane bagasse (Yang et al. 2012) and also pretreatment of birch (Goshadrou et al. 2013). Lennartsson et al. (2011) studied concentrated NMMO pretreatment (85 % w/w, 130 °C, 5 h) of spruce and birch in pilot scale. For wood chips below 2 mm, maximum hydrolysis yields (mg/g wood) ranged from 195 to 128 for spruce and 136 to 175 for birch depending on the scale of the pilot reactor using nonisothermal SSF. A technoeconomic analysis of NMMO pretreatment of spruce for ethanol and biogas production was conducted by Shafiei et al. (2011). They observed relatively high process energy efficiency of 79 %.

4.1.4.4 Urea/Sodium Hydroxide

The Urea/Sodium hydroxide solutions have been found to dissolve cellulose at a subzero temperature for the homogeneous synthesis of cellulose derivatives (Khodaverdi et al. 2012; Wang et al. 2008, 2011a; Ruan et al. 2004). Khodaverdi et al. (2012) used urea/sodium hydroxide solution to pretreat spruce. Spruce showed slight removal of cellulose, hemicellulose, and lignin. However, a significant increase in enzymatic glucan digestibility was observed (Khodaverdi et al. 2012). But, it may be too expensive to prepare prechilled urea/sodium hydroxide and recycle this solution, particularly in the case of biomass pretreatment that is used to produce low-value biocommodities. For example, sodium hydroxide-based pulping used to cause serious water pollution in China. So it has been abandoned (Sathitsuksanoh et al. 2012a). It should be noted that pulp is several times more valuable than ethanol.

4.1.4.5 *N,N*-Dimethylacetamide (DMAc)/LiCl

N,N-Dimethylacetamide (DMAc)/LiCl solution can dissolve cellulose (Cai et al. 2004) because hydrogen bonding of the hydroxyl protons of cellulose with the chloride ions allows the solvent to penetrate into cellulose fibers. DMAc/LiCl is suitable for processing and derivatizing pure cellulose (Striegel 1997). Wang et al. (2011b) conducted a comparative study using different cellulose solvents—LiOH/

urea, LiCl/DMAc, concentrated phosphoric acid, NMMO, and 1-butyl-3-methylimidazolium chloride. Except for the cellulosic sample regenerated from LiCl/DMAc system, all the other treated samples showed reduced cellulose crystallinity and the degree of polymerization and as a consequence, showed a significant improvement of enzymatic hydrolysis kinetics. The regenerated cellulose from concentrated phosphoric acid almost completely consisted of cellulose II, and obtained the highest saccharification yield (Sathitsuksanoh et al. 2012a).

4.1.5 Biological Treatment

Biological pretreatment uses wood degrading microorganisms which include white-rot fungi, brown-rot fungi, soft-rot fungi, and bacteria to modify the structure/chemical composition of the lignocellulosic biomass so that the modified biomass is more responsive to enzymatic digestion. Fungi have distinct degradation characteristics on lignocellulosic biomass. Generally, brown and soft rots mainly attack cellulose and impart minor modifications to lignin, whereas white-rot fungi more actively degrade the lignin component. Currently, research is aimed toward finding those organisms which can degrade lignin more effectively and more specifically. White rot fungi are considered the most promising basidiomycetes for bio-pretreatment of biomass and are the most studied biomass degrading microorganisms (Lee 1997; Sun and Cheng 2002). Brown-rot, white, and soft-rot fungi attack wood via the production of enzymes such as lignin peroxidases, polyphenol oxidases, manganese-dependent peroxidases, and laccases that degrade the lignin. Hatakka et al. (1993) reported the selective delignification of wood and wheat straw by selected white-rot fungi such as; *Phanerochaete chrysosporium*, *Phlebia radiata*, *Dichmitus squalens*, *Rigidosporus lignosus*, and *Jungua separabilima*. Lignin depolymerization by these fungi takes several weeks to achieve significant result but can be very selective and efficient (Hatakka 1994; Hatakka et al. 1993).

Hwang et al. (2008) investigated biological pretreatment of wood chips using four different white-rot fungi for 30 days. They observed that the glucose yield of pretreated wood by *Trametes versicolor* MrP 1 reached 45 % by enzymatic hydrolysis whereas 35 % solid was converted to glucose during fungal incubation. A Japanese red pine *Pinus densiflora* (softwood) was pretreated biologically by several white-rot fungi—*Ceriporia lacerata*, *Stereum hirsutum*, and *Polyporus brumalis* and it was found that *S. hirsutum* is the most effective to degrade lignin and improve the enzymatic digestibility of wood (Lee et al. 2007).

Preliminary studies conducted by Keller et al. (2003) showed a 3- to 5-fold improvement in enzymatic digestibility of corn stover after pretreatment with *Cyathus stercoreus*; and a 10- to 100-fold reduction in shear force required to obtain the same shear rate of 3.2–7.0 rev/s, respectively, after pretreatment with *Phanerochaete chrysosporium*.

Zhang et al. (2007a, b, c) screened 35 isolates of white-rot fungi for the biological pretreatment of bamboo for enzymatic saccharification. They observed that

Echinodontium taxodii 2538, *Trametes versicolor* G20 and *Coriolus versicolor* B1 were the most promising white-rot fungi for significant improvement of enzymatic saccharification and highly selective lignin degradation. Degradation of lignin by white-rot fungi is a cooxidative process. Therefore, accompanying carbon source is necessary, usually from cellulose and hemicellulose. In order to reduce and avoid the sugar loss during bio-pretreatment, the fungal strains with preference of lignin degradation such as *C. subvermispora, Cyathus stercoreus, P. ostreatus* and genetically modified *P. chrysosporium* were developed to produce less cellulase activity (Kerem et al. 1992; Kirk et al. 1986). The biological pretreatment appears to be a promising technique and has very evident advantages, including no chemical requirement, low energy input, mild environmental conditions, low energy input, and environmentally friendly working manner (Sun and Cheng 2002). However, biological pretreatment is very slow and large amount of space is required to perform this treatment. It also requires careful control of growth conditions (Chandra et al. 2007). In addition, most lignolytic fungi solubilize/consume not only lignin but also hemicellulose and cellulose (Lee et al. 2007; Singh et al. 2008; Kuhar et al. 2008; Shi et al. 2008). Therefore, the biological pretreatment faces technoeconomic challenges and is less attractive commercially.

4.1.6 Other Methods

4.1.6.1 Glycerol

Crude glycerol has been also studied as a solvent for fractionating biomass in order to improve the economics of cellulosic ethanol and also upstream biodiesel production. Glycerol pretreatment causes very effective delignification of biomass. Guragain et al. (2011) found the optimum conditions for the use of crude glycerol (water:glycerol = 1:1) in pretreating wheat straw and water hyacinth. Best results were obtained at a temperature of 230 °C for 4 h for wheat straw and 230 °C for 1 h for water hyacinth. Enzymatic hydrolysis of pretreated wheat straw produced reducing sugar yields (mg/g of sample) of 423 and 487 for crude and pure glycerol, respectively, compared to 223 for dilute acid. In addition, hydrolysis tests on water hyacinth gave yields of 705, 719, and 714 for crude glycerol, pure glycerol, and dilute acid, respectively.

Ungurean et al. (2011) conducted glycerol pretreatment of different types of wood—poplar, acacia, oak, and fir—and obtained higher cellulose conversion rates compared to dilute acid application. However, combinations of glycerol and acid/IL pretreatment produced higher sugar levels compared to glycerol pretreatment alone. There are wide variations in the composition of crude glycerol which usually contains methanol, ash, soap, catalysts, salts, and nonglycerol organic matter, among others, in different proportions (Yang et al. 2012). While the potential for exploring crude glycerol application together with other methods is high, there is a need to assess the quality of crude glycerol and its effects on sugar and ethanol yields of promising feedstocks.

4.1.6.2 Wet Oxidation

Wet oxidation process was originally used as a means of wastewater treatment and soil remediation, and later on it was also used in the pretreatment of lignocellulosic feedstocks (Chaturvedi and Verma 2013). Wet oxidation involves water and air, oxygen, or hydrogen peroxide at high pressure and temperature (Varga et al. 2003). Most of the hemicellulose is dissolved in the pretreatment, and it is possible to obtain a moderate to high degree of delignification through oxidation (controlled combustion). In this process, large number of organic polymers mostly hemicellulose and lignin are converted to oxidized compounds, such as low-molecular-weight carboxylic acids, alcohol, or even carbon dioxide and water. Wet oxidization is mostly carried out with the addition of an alkali such as sodium carbonate to reduce the reaction temperature and the amount of hemicellulose being oxidized. The use of hydrogen peroxide has attracted much interest in the recent years, but the high cost of this chemical could make this technology economically prohibitive. The introduction of pure oxygen to the reaction has been challenged, since uncontrolled combustion can occur at the oxygen injection points. Therefore, it is very much unlikely that this pretreatment method will find practical applications in biomass processing.

In the wet oxidation process, oxidation of organic matter in the presence of oxygen takes place. When the process takes place at low temperatures, hydrolysis of lignocellulose occurs. At high temperatures, oxidation of lignocellulose occurs with liberation of carbon dioxide and water. Wet oxidation process solubilizes hemicellulose and lignin is degraded into carbon dioxide, water, and carboxylic acids (succinic acid, formic acid, acetic acid, glycolic acid) and phenolic compounds (Bjerre et al. 1996). It was presented as an alternative to steam explosion process which had become the most widely used pretreatment method (Palonen et al. 2004). Industrially, wet air oxidation processes have been used for the treatment of wastes with a high organic matter by oxidation of soluble or suspended materials using either an oxidizing agent such as hydrogen peroxide or oxygen in aqueous phase at high temperatures (180–200 °C) (Jorgensen et al. 2007).

The effect of wet oxidation to improve anaerobic biodegradability and methane yields from different biowastes such as food waste, yard waste, and digested biowaste by using thermal wet oxidation was studied by Lissens et al. (2004). Methane yields for wet oxidized yard waste, raw food waste, wet oxidized food, and raw yard waste were 345, 685, 536, 571, and 345 mL of methane/g of volatile suspended solids, respectively. Higher oxygen pressure during wet oxidation of digested biowaste substantially increased the total methane yield and digestion kinetics and also allowed lignin utilization during a subsequent second digestion. The increase of the specific methane yield for the full-scale biogas plant by using thermal wet oxidation was 35–40 %, showing that there is still a significant amount of methane that can be harvested from anaerobic digested biowaste.

Szijártó et al. (2009) treated common reed (*Phragmites australis*) by using the wet oxidation process to improve the enzymatic digestibility of reed cellulose to soluble sugars, thus improving the convertibility of reed to ethanol. The most effective treatment increased the digestibility of reed cellulose by cellulase enzymes more than three times compared to the untreated control. About 51.7 % of the hemicellulose and 58.3 % of the lignin was solubilized and about 87.1 % of the cellulose remained in the solids during this wet oxidation process. The conversion of cellulose to glucose was 82.4 % after enzymatic hydrolysis of pretreated fibers from the same treatment. Simultaneous saccharification and fermentation of pretreated solids resulted in a final ethanol concentration as high as 8.7 g/L which was 73 % of the theoretical yield.

Pretreatment of rice husk by wet air oxidation for the production of ethanol was studied by Banerjee et al. (2009). Optimum conditions were found to be 0.5 MPa, 185 °C, and 15 min. About 67 % (w/w) cellulose content in the solid fraction was obtained with 89 % lignin removal, and 70 % hemicellulose solubilization.

Banerjee et al. (2011) also investigated pretreatment of rice husk by Alkaline Peroxide—Assisted Wet Air Oxidation (APAWAO) to increase the enzymatic convertibility of cellulose in pretreated rice husk. Rice husk was presoaked overnight in 1 % (w/v) hydrogen peroxide solution at room temperature, followed by wet air oxidation. APAWAO pretreatment resulted in solubilization of 67 wt% of hemicellulose and 88 wt% of lignin initially present in raw rice husk. APAWAO pretreatment resulted in 13-fold increase in the amount of glucose that could be obtained from otherwise untreated rice husk. Up to 86 % of cellulose in the pretreated rice husk could be converted into glucose within 24 h, yielding over 21 % glucose. For pretreatment of biomass having high lignin content, wet oxidation process is considered suitable. This process has shown promising results with different types of biomass. Reducing sugar yields up to 70 % have been obtained by utilizing this pretreatment process (Banerjee et al. 2009). The disadvantage of this process is that it requires high temperature, pressure, and the presence of strong oxidizing agents such as hydrogen peroxide. This lead to high costs to maintain such conditions and also require large-scale reaction vessels to tolerate such harsh conditions. Therefore, application of this process in large-scale pretreatment of biomass is not very common.

In the wet oxidation process, lower amounts of furfural and 5-hydroxymethylfurfural are produced. These are strong inhibitors in the fermentation step, as a result of oxygen degradation of these components. However, for the same reason, the large amount of hemicellulose sugars are lost, and therefore the overall process yield would be reduced and the economy is not attractive. Another concern associated with the oxidation process is its exothermal nature, which requires careful control of the process parameters.

4.1.6.3 Microwaves Pretreatment

Microwaves have been used in the treatment of lignocellulosic biomass (Chaturvedi and Verma 2013). Microwaves cause localized heating of biomass which leads to disruption of lignocellulose architecture. Thus, cellulose and hemicellulose become accessible to enzymatic hydrolysis (Sarkar et al. 2012). Microwave chemistry offers many advantages over conventional methods of heating. The process is more energy efficient because microwave irradiation heats the whole volume of a sample whereas conventional heating heats the sample in contact with the reaction vessel before the bulk. The heating effect is almost rapid, unlike conventional heating methods. There is no time spent waiting for the source to heat up or cool down. Microwave reactor operating systems allow easy control of temperature and pressure to exact and steady values. The speed of microwave systems also allows for quick screening of parameters and evaluation of new methods (Kappe 2004). The application of microwave technology in high throughput reactions such as flow reactors adds further to the advantages of microwave irradiation (Roberts and Strauss 2005; Wilson et al. 2004).

Su et al. (2010) studied the effects of microwave treatment on sorghum liquor waste for bioethanol production. Their results showed that reducing sugar yield following microwave treatment was very high as compared to untreated waste. Liu et al. (2010) studied microwave pretreatment of recalcitrant softwood in the presence of aqueous glycerol and different organic and inorganic acids. They found that the pulp obtained by organosolvlysis with 0.1 % hydrochloric acid (pKa-6) at 180 °C for 6 min gave the highest sugar yield, 53.1 %. With other acids such as malonic acid, and phosphoric acid having lower pKa value, a lower sugar yield was obtained. Microwave-assisted glycerolysis is a suitable process for treatment of different types of soft woods.

Maa et al. (2009) studied microwave pretreatment of rice straw. Under optimized conditions consisting of microwave intensity 680 W, substrate concentration 75 g/L, irradiation time 24 min, maximal efficiencies of cellulose, hemicellulose and total saccharification were, respectively, increased by 30.6, 43.3, and 30.3 %. Microwave treatment of biomass is a harsh process. It leads to increased sugar yields and high lignin removal. Studies have shown that the yield of reducing sugars ranges from 40 to 60 % (Verma et al. 2011). Promising results are obtained by microwave pretreatment of biomass. However, the process is very expensive when efficacy of this process in terms of cost is considered (Feng and Chen 2008). A large microwave irradiator is required for large-scale pretreatment of biomass. The process is energy consuming and time consuming and limits its use in large-scale operations. Another major drawback of the process is the generation of high temperature and nonuniform heating of biomass. This leads to the generation of inhibitors and therefore the yields are generally lower as expected. The operational cost is also increased (Jackowiak et al. 2011).

References

Agbor VB, Cicek N, Sparling R, Berlin A, Levin DB (2011) Biomass pretreatment: fundamentals toward application. Biotechnol Adv 29:675–685

Alizadeh H, Teymouri F, Gilbert TI, Dale BE (2005) Pretreatment of switchgrass by ammonia fibre explosion (AFEX). Appl Biochem Biotechnol 121–124:1133–1141

Alriols MG, Tejado A, Blanco M, Mondragon I, Labidi J (2009) Agricultural palm oil tree residues as raw material for cellulose, lignin and hemicelluloses production by ethylene glycol pulping process. Chem Eng J 148(1):106–114
Alriols MG, Garcıa A, Llano-ponte R, Labidi J (2010) Combined organosolv and ultrafiltration lignocellulosic biorefinery process. Chem Eng J 157(1):113–120
Alvira P, Tomás-Pejó E, Ballesteros M, Negro MJ (2010) Pretreatment technologies for an efficient bioethanol production process based on enzymatic hydrolysis: a review. Bioresour Technol 101(13):4851–4861
Ang TN, Ngoh GC, Chua ASM, Lee MG (2012) Elucidation of the effect of ionic liquid pretreatment on rice husk via structural analyses. Biotechnol Biofuels 5, article 67
Antal MJ Jr (1996) Water: a traditional solvent pregnant with new application. In: White HJ Jr (ed) Proceedings of the 12th international conference on the properties of water and steam. Begell House, New York, p 23–32
Arato C, Pye EK, Gjennestad G (2005) The lignol approach to biorefining of woody biomass to produce ethanol and chemicals. Appl Biochem Biotechnol A 123(1–3):871–882
Avella R, Scoditti E (1998) The Italian steam explosion program of ENEA. Biomass Bioenergy 14:289–293
Bak JS, Ko JK, Han YH (2009) Improved enzymatic hydrolysis yield of rice straw using electron beam irradiation pretreatment. Bioresour Technol 100:1285–1290
Banerjee S, Sen R, Pandey RA, Chakrabarti T, Satpute D, Giri BS, Mudliar S (2009) Evaluation of wet air oxidation as a pretreatment strategy for bioethanol production from rice husk and process optimization. Biomass Bioenergy 33:1680–1686
Banerjee S, Sen R, Mudliar S, Pandey RA, Chakrabarti T, Satpute D (2011) Alkaline peroxide assisted wet air oxidation pretreatment approach to enhance enzymatic convertibility of rice husk. Biotechnol Prog 27(3):691–697
Banerjee G, Car S, Liu T (2012) Scale-up and integration of alkaline hydrogen peroxide pretreatment, enzymatic hydrolysis, and ethanolic fermentation. Biotechnol Bioeng 109 (4):922–931
Ben-Ghedalia D, Miron J (1981) The effect of combined chemical and enzyme treatment on the saccharification and in vitro digestion rate of wheat straw. Biotechnol Bioeng 23:823–831
Bensah EC, Mensah M (2013) Chemical pretreatment methods for the production of cellulosic ethanol: technologies and innovations. Int J Chem Eng 2013:21 p, Article ID719607
Bjerre AB, Olesen AB, Fernqvist T (1996) Pretreatment of wheat straw using combined wet oxidation and alkaline hydrolysis resulting in convertible cellulose and hemicellulose. Biotechnol Bioeng 49:568–577
Bobleter O (1994) Hydrothermal degradation of polymers derived from plants. Prog Polym Sci 19:797–841
Brennan AH, Hoagland W, Schell DJ (1986) High temperature acid hydrolysis of biomass using an engineering-scale plug flow reactor: result of low solids testing. Biotechnol Bioeng Symp 17:53–70
Brink DL (1994) Method of treating biomass material. US Patent 5:366–558
Brosse N, Hage REL, Sannigrahi P, Ragauskas A (2010) Dilute sulphuric acid and ethanol organosolv pretreatment of miscanthus x giganteus. Cellul Chem Technol 44(1–3):71–78
Bu L, Xing Y, Yu H, Gao Y, Jiang J (2012) Comparative study of sulfite pretreatments for robust enzymatic saccharification of corn cob residue. Biotechnol Biofuels 5, article 8
Cadoche L, López GD (1989) Assessment of size reduction as a preliminary step in the production of ethanol from lignocellulosic wastes. Biotechnol Wastes 30:153–157
Cahela DR, Lee YY, Chambers RP (1983) Modeling of percolation process in hemicellulose hydrolysis. Biotechnol Bioeng 25:3–17
Cai J, Zhang L, Zhou J, Li H, Chen H, Jin H (2004) Novel fibers prepared from cellulose in NaOH/urea aqueous solution. Macromol Rapid Commun 25:1558–1562
Cao Y, Li H, Zhang Y, Zhang J, He J (2010) Structure and properties of novel regenerated cellulose films prepared from cornhusk cellulose in room temperature ionic liquids. J Appl Polym Sci 116:547–554

Cao W, Sun C, Liu R, Yin R, Wu X (2012) Comparison of the effects of five pretreatment methods on enhancing the enzymatic digestibility and ethanol production from sweet sorghum bagasse. Bioresour Technol 111:215–221

Chan YL, Yang J, Ahn JW, Moon YH, Yoon YM, Gyeong-Dan Yu GD, An GH, Choi IH (2014) The optimized CO_2-added ammonia explosion pretreatment for bioethanol production from rice straw. Bioprocess Biosyst Eng 37(9):1907–1915

Chandra RP, Bura R, Mabee WE (2007) Substrate pretreatment: the key to effective enzymatic hydrolysis of lignocellulosics? Adv Biochem Engin/Biotechnol 108:67–93

Chang VS, Holtzapple MT (2000) Fundamental factors affecting biomass enzymatic reactivity. Appl Biochem Biotechnol 84–86:5–37

Chang MM, Chou TYC, Tsao GT (1981) Structure, pretreatment and hydrolysis of cellulose. Adv Biochem Eng 20:15–42

Chang VS, Burr B, Holtzapple MT (1997) Lime pretreatment of switchgrass. Appl Biochem Biotechnol 63–65:3–19

Chang VS, Nagwani M, Holtzapple MT (1998) Lime pretreatment of crop residues bagasse and wheat straw. Appl Biochem Biotechnol 74:135–159

Chang VS, Nagwani M, Kim CH, Holtzapple MT (2001) Oxidative lime pretreatment of high-lignin biomass. Appl Biochem Biotechnol 94:1–28

Chaturvedi V, Verma P (2013) An overview of key pretreatment processes employed for bioconversion of lignocellulosic biomass into biofuels and value added products. 3 Biotech 3:415–431. doi:10.1007/s13205-013-0167-8

Cheng KK, Wang W, Zhang JA, Zhao Q, Li JP, Xue JW (2011) Statistical optimization of sulfite pretreatment of corncob residues for high concentration ethanol production. Bioresour Technol 102(3):3014–3019

Chosdu R, Hilmy N, Erizal Erlinda TB, Abbas B (1993) Radiation and chemical pretreatment of cellulosic waste. Radiat Phys Chem 42:695–698

Chum HL, Douglas LJ, Feinberg DA, Schroeder HA (1985) Evaluation of pretreatments of biomass for enzymatic hydrolysis of cellulose. US Department of Energy, Contract No. DE-AC02-83CHt0093

Conde-Mejía C, Jiménez-Gutiérrez A, El-Halwagi M (2012) A comparison of pretreatment methods for bioethanol production from lignocellulosic materials. Saf Environ Prot 90(3):189–202

Conte P, Maccotta A, Pasquale CD, Bubici S, Alonzo G (2009) Dissolution mechanism of crystalline cellulose in H_3PO_4 as assessed by high field NMR spectroscopy and fast field cycling NMR relaxometry. J Agric Food Chem 57:8748–8752

Converse AO, Kwarteng IK, Grethlein HE, Ooshima H (1989) Kinetics of thermochemical pretreatment of lignocellulosic materials. Appl Biochem Biotechnol 20/21:63–78

Correia JA, Júnior JE, Gonçalves LR, Rocha MV (2013) Alkaline hydrogen peroxide pretreatment of cashew apple bagasse for ethanol production: study of parameters. Bioresour Technol 139:249–256

Dadi AP, Schall CA, Varanasi S (2006) Enhancement of cellulose saccharification kinetics using an ionic liquid pretreatment step. Biotechnol Bioeng 95:904–910

Dadi AP, Schall CA, Varanasi S (2007) Mitigation of cellulose recalcitrance to enzymatic hydrolysis by ionic liquid pretreatment. Appl Biochem Biotechnol 137–140(1–12):407–421

Dale BE, Moreira MJ (1982) A freeze-explosion technique for increasing cellulose hydrolysis. Biotechnol Bioeng Symp 12:31–43

Dale BE, Henk LL, Shiang M (1984) Fermentation of lignocellulose materials treated my ammonia freeze-explosion. Dev Ind Microbiol 26:223–233

Damaso MCT, Mde Castro, Castro RM, Andrade MC, Pereira N (2004) Application of xylanase from Thermomyces lanuginosus IOC-4145 for enzymatic hydrolysis of corn cob and sugarcane bagasse. Appl Biochem Biotechnol 115:1003–1012

Delgenes JP, Penaud V, Moletta R (2002) Pretreatment for the enhancement of anaerobic digestion of solid waster Chap. 8. In: Biomethanization of the organic fraction of municipal solid waste. IWA publishing, pp 201–228

Duff SJB, Murray WD (1996) Bioconversion of forest products industry waste cellulosics to fuel ethanol: a review. Bioresour Technol 55:1–33
Dunlap CE, Chiang LC (1980) Cellulose degradation-a common link. In: Shuler ML (ed) Utilization and recycle of agricultural wastes and residues. CRC Press, Boca Raton, FL, pp 19-65
Eggeman T, Elander TR (2005) Process and economic analysis of pretreatment technologies. Biores Technol 8:2019–2025
Esteghalian A, Hashimoto AG, Fenske JJ, Penner MH (1997) Modelling and optimization of dilute-sulfuric-acid pretreatment of corn stove, poplar and switchgrass. Bioresour Technol 59:129–136
Euphrosine-Moy V, Lasry T, Bes RS, Molinier J, Mathieu J (1991) Degradation of poplar lignin with ozone. Ozone Sci Eng 13(2):239–248
Fan LT, Gharpuray MM, Lee YH (1987) Cellulose hydrolysis biotechnology monographs. Springer, Berlin, p 57
Feng L, Chen Z (2008) Research progress on dissolution and functional modification of cellulose in ionic liquids. J Mol Liq 142:1–5
Fink HP, Weigel P, Purz HJ (2001) Structure formation of regenerated cellulose materials from NMMO-solutions. J Prog Polym Sci 26:1473–1524
Foster BL, Dale BE, Peterson JBD (2001) Enzymatic hydrolysis of ammonia-treated sugar beet pulp. Appl Biochem Biotechnol 91–93:269–282
Galbe M, Zacchi G (2007) Pretreatment of lignocellulosic materials for efficient bioethanol production. Biofuels 108:41–65
Ge X, Green VS, Zhang N, Sivakumar G, Xu J (2012) Eastern gamagrass as an alternative cellulosic feedstock for bioethanol production. Process Biochem 47:335–339
Geng A, Xin F, Ip J (2012) Ethanol production from horticultural waste treated by a modified organosolv method. Bioresour Technol 104:715–721
Goshadrou A, Karimi K, Taherzadeh MJ (2013) Ethanol and biogas production from birch by NMMO pretreatment. Biomass Bioenergy 49:95–101
Gregg DJ, Saddler JN (1996) Factors affecting cellulose hydrolysis and the potential of enzyme recycle to enhance the efficiency of an integrated wood to ethanol process. Biotechnol Bioeng 51:375–383
Grous WR, Converse AO, Grethlein HE (1986) Effect of steam explosion pretreatment on pore size and enzymatic hydrolysis of poplar. Enzym Microb Technol 8:274–280
Guragain YN, De Coninck J, Husson F, Durand A, Rakshit SK (2011) Comparison of some new pretreatment methods for second generation bioethanol production from wheat straw and water hyacinth. Bioresour Technol 102(6):4416–4424
Hammel KE, Kapich AN, Jensen KA Jr, Ryan ZC (2002) Reactive oxygen species as agents of wood decay by fungi. Enzym Microb Technol 30(4):445–453
Hatakka A (1994) Lignin-modifying enzymes from selected white-rot fungi: production and role from in lignin degradation. FEMS Microbiol Rev 13:125–135
Hatakka AI, Varesa T, Lunn TK (1993) Production of multiple lignin peroxidases by the white-rot fungus Phlebia ochraceofulva. Enzym Microb Technol 15:664–669
Haverty D, Dussan K, Piterina AV, Leahy JJ, Hayes MHB (2012) Autothermal, single-stage, performic acid pretreatment of Miscanthus x giganteus for the rapid fractionation of its biomass components into a lignin/hemicellulose-rich liquor and a cellulase-digestible pulp. Bioresour Technol 109:173–177
Heinze T, Schwikal K, Barthel S (2005) Ionic liquids as reaction medium in cellulose functionalization. Macromol Biosci 5:520–525
Hendricks AT, Zeeman G (2009) Pretreatments to enhance the digestibility of lignocellulosic biomass. Bioresour Technol 100:10–18
Hideno A, Kawashima A, Endo T, Honda K, Morita M (2013) Ethanol-based organosolv treatment with trace hydrochloric acid improves the enzymatic digestibility of Japanese cypress (Chamaecyparis obtusa) by exposing nanofibers on the surface. Bioresour Technol 18:64–70

Hinman ND, Schell DJ, Riley CJ, Bergeron PW, Walter PJ (1992) Preliminary estimate of the cost of ethanol production for SSF technology. Appl Biochem Biotechnol 34/35:639–649
Holm J, Lassi U (2011) Ionic liquids in pretreatment of lignocellulosic biomass. In: Kokorin A (ed) Ionic liquids: application and perspectives. In-Tech, pp 546–560
Holtzapple MT, Jun JH, Ashok G, Patibandla SL, Dale BE (1990) Ammonia fiber explosion (AFEX) pretreatment of lignocellulosic wastes. In: American Institute of Chemical Engineers National Meeting, Chicago, IL
Holtzapple MT, Jun JH, Ashok G, Patibandla SL, Dale BE (1991) The ammonia freeze explosion (AFEX) process: a practical lignocellulose pretreatment. Appl Biochem Biotechnol 28/29: 59–74
Holtzapple MT, Lundeen JE, Sturgis R, Lewis JE, Dale BE (1992a) Pretreatment of lignocellulosic municipal solid-waste by ammonia fibre explosion (AFEX). Appl Biochem Biotechnol 34–35:5–21
Holtzapple MT, Davison RR, Stuart ED (1992b) Biomass refining process. US Patent 5(171):592
Hou XD, Smith TJ, Li N, Zong MH (2012) Novel renewable ionic liquids as highly effective solvents for pretreatment of rice straw biomass by selective removal of lignin. Biotechnol Bioeng 109(10):2484–2493
Hu Z, Wen Z (2008) Enhancing enzymatic digestibility of switchgrass by microwave-assisted alkali pretreatment. Biochem Eng J 38:369–378
Hu Z, Wang Y, Wen Z (2008) Alkali (NaOH) pretreatment of switchgrass by radio frequency-based dielectric heating. Appl Biochem Biotechnol 148:71–81
Huijgen WJJ, Van der Laan RR, Reith JH (2008) Modified organosolv as a fractionation process of lignocellulosic biomass for coproduction of fuels and chemicals. In: Proceedings of the 16th European biomass conference and exhibition, Valencia, Spain
Hwang SS, Lee SJ, Kim HK (2008) Biodegradation and saccharification of wood chips of Pinus strobus and Liriodendron tulipifera by white rot fungi. J Microbiol Biotechnol 18:1819–1825
Ichwan M, Son TW (2011) Study on organosolv pulping methods of oil palm biomass. In: International seminar on chemistry, pp 364–370
Imai M, Ikari K, Suzuki I (2004) High-performance hydrolysis of cellulose using mixed cellulase species and ultrasonication pretreatment. Biochem Eng J 17:79–83
Isci A, Himmelsbach JN, Pometto AL, Raman R, Anex RP (2008) Aqueous ammonia soaking of switchgrass followed by simultaneous saccharification and fermentation. Appl Biochem Biotechnol 144:69–77
Itoh H, Wada M, Honda Y, Kuwahara M (2003) Bioorganosolve pretreatments for simultaneous saccharification and fermentation of beech wood by ethanolysis and white-rot fungi. J Biotechnol 103(3):273–280
Iyer PV, Wu ZW, Kim SB, Lee YY (1996) Ammonia recycled percolation process for pretreatment of herbaceous biomass. Appl Biochem Biotechnol 57–58:121–132
Jackowiak D, Bassard D, Pauss A, Ribeiro T (2011) Optimisation of a microwave pretreatment of wheat straw for methane production. Bioresour Technol 102(12):6750–6756
Jin Y, Yang L, Jameel H, Chang HM, Phillips R (2013) Sodium sulfite-formaldehyde pretreatment of mixed hardwoods and its effect on enzymatic hydrolysis. Bioresour Technol 135:109–115
Jorgensen H, Kristensen JB, Felby C (2007) Enzymatic conversion of lignocellulose into fermentable sugars: challenges and opportunities. Biofuels Bioprod Bioref 1:119–134
Kappe CO (2004) Controlled microwave heating in modern organic synthesis. Angew Chem Int Ed 43:6250–6284
Karr WE, Holtzapple MT (1998) The multiple benefits of adding non-ionic surfactant during the enzymatic hydrolysis of corn stover. Biotechnol Bioeng 59:419–427
Karr WE, Holtzapple T (2000) Using lime pretreatment to facilitate the enzymatic hydrolysis of corn stover. Biomass Bioenergy 18:189–199
Keller FA, Hamilton JE, Nguyen QA (2003) Microbial pretreatment of biomass-potential for reducing severity of thermochemical biomass pretreatment. Appl Biochem Biotechnol 105: 27–41

Kerem Z, Friesem D, Hadar Y (1992) Lignocellulose degradation during solid-state fermentation Pleurotus ostreatus versus Phanerochaete chrysosporium. Appl Environ Microbiol 58:1121–1127

Khodaverdi M, Jeihanipour A, Karimi K, Taherzadeh MJ (2012) Kinetic modeling of rapid enzymatic hydrolysis of crystalline cellulose after pretreatment by NMMO. J Ind Microbiol Biotechnol 39:429–438

Kilzer FJ, Broido A (1965) Speculations on the nature of cellulose pyrolysis. Pyrodynamics 2:151–163

Kim S, Holtzapple MT (2005) Lime pretreatment and enzymatic hydrolysis of corn stover. Bioresour Technol 96:1994–2006

Kim S, Holtzapple MT (2006a) Effect of structural features on enzyme digestibility of corn stover. Bioresour Technol 97:583–591

Kim S, Holtzapple MT (2006b) Lime pretreatment and enzymatic hydrolysis of corn stover. Bioresour Technol 96:1994–2006

Kim HK, Hong J (2001) Supercritical CO_2 pretreatment of lignocellulose enhances enzymatic cellulose hydrolysis. Bioresour Technol 77:139–144

Kim SB, Lee YY (2002) Diffusion of sulfuric acid within lignocellulosic biomass particles and its impact on dilute-acid pretreatment. Bioresour Technol 83:165–171

Kim TH, Lee YY (2005a) Pretreatment and factionation of corn stover by soaking in aqueous ammonia. Appl Biochem Biotechnol 121:1119–1131

Kim TH, Lee YY (2005b) Pretreatment of corn stover by ammonia recycle percolation process. Bioresour Technol 96:2007–2013

Kim SB, Yum DM, Park SC (2000) Step-change variation of acid concentration in a percolationreactor for hydrolysis of hardwood hemicellulose. Bioresour Technol 72:289–294

Kim HT, Kim JS, Sunwoo C, Lee YY (2003) Pretreatment of corn stover by aqueous ammonia. Biores Technol 90:39–47

Kim TH, Lee YL, Sunwoo C, Kim JS (2006) Pretreatment of corn stover by low-liquid ammonia recycle percolation process. Appl Biochem Biotechnol 133:41–57

Kim Y, Yu A, Han M, Choi GW, Chung B (2011) Enhanced enzymatic saccharification of barley strawpretreated by ethanosolv technology. Appl Biochem Biotechnol 163(1):143–152

Kim HY, Gwak KS, Lee SY, Jeong HS, Ryu KO, Choi IG (2012) Biomass characteristics and ethanol production of yellow poplar (Liriodendron tulipifera) treated with slurry composting and biofiltration liquid as fertilizer. Biomass Bioenergy 42:10–17

Kirk TK, Tien M, Johnsrud SC (1986) Lignin degrading activity of Phanerochaete chrysosporium burds: comparision of cellulase-negative and other strains. Enzym Microb Technol 8:75–80

Kitchaiya P, Intanakul P, Krairiksh MJ (2003) Enhancement of enzymatic hydrolysis of lignocellulosic wastes by microwave pretreatment under atmospheric-pressure. J Wood Chem Technol 23:217–225

Kohlmann KL, Sariyaka A, Westgate PJ, Weil J, Velayudhan A, Hendrickson R, Ladisch MR (1995) Enhance enzyme activities on hydrated lignocellulosic substrates. In: Saddler JN, Penner MH (eds) Enzymatic degradation of insoluble carbohydrates. ACS publishing, pp 237–255

Kong F, Engler CR, Soltes EJ (1992) Effects of cell-wall acetate, xylan backbone, and lignin on enzymatic hydrolysis of aspen wood. Appl Biochem Biotechnol 34(35):23–35

Koo BW, Min BC, Gwak KS (2012) Structural changes in lignin during organosolv pretreatment of Liriodendron tulipifera and the effect on enzymatic hydrolysis. Biomass Bioenergy 42:24–32

Kuhad RC, Singh A, Ericksson KE (1997) Microorganisms and enzymes involved in the degradation of plant fiber cell walls. Adv Biochem Eng Biotechnol 57:45–125

Kuhar S, Nair LM, Kuhad RC (2008) Pretreatment of lignocellulosic material with fungi capable of higher lignin degradation and lower carbohydrate degradation improves substrate acid hydrolysis and the eventual conversion to ethanol. Can J Microbiol 54:305–313

Kumar P, Barrett DM, Delwiche MJ, Stroeve P (2009) Methods for pretreatment of lignocellulosic biomass for efficient hydrolysis and biofuel production. Ind Eng Chem Res 48:3713–3729

Kuo CH, Lee CK (2009a) Enhanced enzymatic hydrolysis of sugarcane bagasse by N-methylmorpholine-N-oxide pretreatment. Bioresour Technol 100:866–871

Kuo CH, Lee CK (2009b) Enhancement of enzymatic saccharification of cellulose by cellulose dissolution pretreatments. Carbohydr Polym 77:41–46

Kupiainen L, Ahola J, Tanskanen J (2012) Hydrolysis of organosolv wheat pulp in formic acid at high temperature for glucose production. Bioresour Technol 116:29–35

Ladisch MR, Ladisch CM, Tsao GT (1978) Cellulose to sugars: new path gives quantitative yield. Science 201:743–745

Lasry T, Laurent JL, Euphrosine-Moy V, Bes RS, Molinier J, Mathieu I (1990) Identification and evaluation of polar sawdust ozonation products. Analysis 18:192–199

Lee J (1997) Biological conversion of lignocellulosic biomass to ethanol. J Biotechnol 56:1–24

Lee YH, Fan LT (1982) Kinetic studies of enzymatic hydrolysis of insoluble cellulose: analysis of the initial rates. Biotechnol Bioeng 24:2383–2406

Lee JW, Gwak KS, Park JY (2007) Biological pretreatment of softwood Pinus densiflora by three white rot fungi. J Microbiol 45:485–491

Lennartsson PR, Niklasson C, Taherzadeh MJ (2011) A pilot study on lignocelluloses to ethanol and fish feed using NMMO pretreatment and cultivation with zygomycetes in an air-lift reactor. Bioresour Technol 102(6):4425–4432

Li B, Asikkala J, Filpponen I, Argyropoulos DS (2010a) Factors affecting wood dissolution and regeneration of ionic liquids. Ind Eng Chem Res 49(5):2477–2484

Li C, Knierim B, Manisseri C (2010b) Comparison of dilute acid and ionic liquid pretreatment of switchgrass: biomass recalcitrance, delignification and enzymatic saccharification. Bioresour Technol 101(13):4900–4906

Li L, Yu ST, Liu FS, Xie CS, Xu CZ (2011) Efficient enzymatic in situ saccharificatio of cellulose in aqeous-ionic liquid media by microwave treatment. BioResources 6(4):4494–4504

Li F, Yao R, Wang H, Hu H, Zhang R (2012a) Process optimization for sugars production from rice straw via pretreatment with sulphur trioxide micro-thermal explosion. BioResources 7 (3):3355–3366

Li Q, Gao Y, Wang H, Li B (2012b) Comparison of different alkali-based pretreatments of corn stover for improving enzymatic saccharification. Bioresour Technol 125:193–199

Li Q, Ji GS, Tang YB, Gu XD, Fei JJ, Jiang HQ (2012c) Ultrasound-assisted compatible in situ hydrolysis of sugarcane bagasse in cellulase-aqueous-N-methylmorpholine-N-oxide system for improved saccharification. Bioresour Technol 107:251–257

Li Z, Jiang Z, Fei B (2012d) Ethanol organosolv pretreatment of bamboo for efficient enzymatic saccharification. BioResources 7(3):3452–3462

Lin KW, Ladisch MR, Schaefer DM (1981) Review on effect of pretreatment on digestibility of cellulosic materials. AIChE Symp 207:102–106

Lissens G, Thomsen AB, De Baere L, Verstraete W, Ahring BK (2004) Thermal wet oxidation improves anaerobic biodegradability of raw and digested biowaste. Environ Sci Technol 38 (12):3418–3424

Liu LY, Chen HZ (2006) Enzymatic hydrolysis of cellulose materials treated with ionic liquid [BMIM]Cl. Chin Sci Bull 51:2432–2436

Liu H, Zhu JY (2010) Eliminating inhibition of enzymatic hydrolysis by lignosulfonate in unwashed sulfite-pretreated aspen using metal salts. Bioresour Technol 101(23):9120–9127

Liu J, Takada R, Karita S, Watanabe T, Honda Y, Watanabe T (2010) Microwave-assisted pretreatment of recalcitrant softwood in aqueous glycerol. Bioresour Technol 23:9355–9360

Lucas M, Hanson SK, Wagner GL, Kimball DB, Rector KD (2012) Evidence for roomtemperature delignification ofwood using hydrogen peroxide and manganese acetate as a catalyst. Bioresour Technol 119:174–180

Lynd LR, Elander RT, Wyman CE (1996) Likely features and costs of mature biomass ethanol technology. Appl Biochem Biotechnol 57–58:741–761

Lynd LR, Jin H, Michels JG, Wyman CE, Dale B (2003) Bioenergy: background, potential, and policy. Center for Strategic and International Studies, Washington, D.C.

Lynd LR, Laser MS, Bransby D, Dale BE, Davison B, Hamilton R, Himmel M, Keller M, McMillan JD, Sheehan J et al (2008) How biotech can transform biofuels. Nat Biotechnol 26 (2):169–172

Maa H, Liu W, Chen X, Wua Y, Yu Z (2009) Enhanced enzymatic saccharification of rice straw by microwave pretreatment. Bioresour Technol 100:1279–1284

McMillan JD (1994a) Pretreatment of lignocellulosic biomass. Enzym Convers Biomass Fuels Prod 566:292–324

McMillan JD (1994b) Pretreatment of lignocellulosic biomass. In: Himmel ME, Baker JO, Overend RP (eds) Enzymatic conversion of biomass for fuels production. American Chemical Society, Washington, D.C., pp 292–324

Menon V, Rao M (2012) Trends in bioconversion of lignocellulose: biofuels, platform chemicals and biorefinery concept. Prog Energy Combust Sci 38(4):522–550

Mesa L, González E, Cara C, González M, Castro E, Mussatto SI (2011) The effect of organosolv pretreatment variables on enzymatic hydrolysis of sugarcane bagasse. Chem Eng J 168 (3):1157–1162

Mes-Hartree M, Dale BE, Craig WK (1988) Comparison of steam and ammonia pretreatment for enzymatic hydrolysis of cellulose. Appl Microbiol Biotechnol 29:462–468

Millet MA, Baker AJ, Satter LD (1976) Physical and chemical pretreatments for enhancing cellulose saccharification. Biotech Bioeng Symp 6:125–153

Mishima D, Tateda M, Ike M, Fujita M (2006) Comparative study on chemical pretreatments to accelerate enzymatic hydrolysis of aquatic macrophyte biomass used in water purification processes. Bioresour Technol 97(16):2166–2172

Mok WSL, Antal MJ Jr (1992) Uncatalyzed solvolysis of whole biomass hemicellulose by hot compresses liquid water. Ind Eng Chem Res 31:1157–1161

Mok WSL, Antal MJ Jr (1994) Biomass fractionation by hot compressed liquid water. In: Bridgewater AV (ed) Advances in thermochemical biomass conversion, vol 2. Blackie Academic and Professional Publishers, New York, pp 1572–1582

Mora-Pale M, Meli L, Doherty TV, Linhardt RJ, Dordick JS (2011) Roomtemperature ionic liquids as emerging solvents for the pretreatment of lignocellulosic biomass. Biotechnol Bioeng 108(6):1229–1245

Morrison WH, Akin DE (1990) Water soluble reaction products from ozonolysis of grasses. J Agric Food Chem 38:678–681

Mosier N, Hendrickson R, Brewer M, Ho N, Sedlak M, Dreshel R (2005a) Industrial scale-up of pH-controlled liquid hot water pretreatment of corn fiber for fuel ethanol production. Appl Biochem Biotechnol 125:77–97

Mosier N, Wyman CE, Dale BE, Elander R, Lee YY, Holtzapple MT (2005b) Features of promising technologies for pretreatment of lignocellulosic biomass. Bioresour Technol 96:673–686

Moultrop JS, Swatloski RP, Moyna G, Rogers RD (2005) High resolution 13C NMR studies of cellulose and cellulose oligomers in ionic liquid solutions. R Soc Chem Chem Commun 12:1557–1559

Moxley GM, Zhu Z, Zhang Y-HP (2008) Efficient sugar release by the cellulose solvent based lignocellulose fractionation technology and enzymatic cellulose hydrolysis. J Agric Food Chem 56:7885–7890

Muhammad N, Man Z, Bustam MA, Mutalib MIA, Wilfred CD, Rafiq S (2011) Dissolution and delignification of bamboo biomass using amino acid-based ionic liquid. Appl Biochem Biotechnol 165(3–4):998–1009

Murnen HK, Balan V, Chundawat SPS, Bals B, Sousa LDC, Dale BE (2007) Optimization of ammonia fiber expansion (AFEX) pretreatment and enzymatic hydrolysis of Miscanthus x giganteus to fermentable sugars. Biotechnol Prog 23:846–850

Nakamura Y, Daidai M, Kobayashi F (2004) Ozonolysis mechanism of lignin model compounds and microbial treatment of organic acids produced. Water Sci Technol 50(3):167–172

Narayanaswamy N, Faik A, Goetz DJ, Gu T (2011) Supercritical carbon dioxide pretreatment of corn stover and switchgrass for lignocellulosic ethanol production. Bioresour Technol 102 (2011):6995–7000

Neely WC (1984) Factors affecting the pretreatment of biomass with gaseous ozone. Biotechnol Bioeng 26:59–65

Nguyen QA, Tucker MP, Keller FA, Eddy FP (2000) Two-stage dilute-acid pretreatment of softwoods. Appl Biochem Biotechnol 84–86:561–575

Ninomiya K, Yamauchi T, Kobayashi M (2013) Cholinium carboxylate ionic liquids for pretreatment of lignocellulosic materials to enhance subsequent enzymatic saccharification. Biochem Eng J 71:25–29

Nitayavardhana S, Rakshit SK, Grewell D (2008) Ultrasound pretreatment of cassava chip slurry to enhance sugar release for subsequent ethanol production. Biotechnol Bioeng 101:487–496

O'Sullivan AC (1996) Cellulose: the structure slowly unravels. Blackie Acad Prof 4:173–207

Palmowski L, Muller J (1999) Influence of the size reduction of organic waste on their anaerobic digestion. In: II international symposium on anaerobic digestion of solid waste, pp 137–144. Barcelona, 15–17 June 1999

Palmqvist E, Hahn-Hägerdal B (2000a) Fermentation of lignocellulosic hydrolysates I: inhibition and detoxification. Bioresour Technol 74:17–24

Palmqvist E, Hahn-Hägerdal B (2000b) Fermentation of lignocellulosic hydrolysates II: inhibitors and mechanisms of inhibition. Bioresour Technol 74:25–33

Palonen H, Thomsen AB, Tenkanen M, Schmidt AS, Viikari L (2004) Evaluation of wet oxidation pretreatment for enzymatic hydrolysis of softwood. Appl Biochem Biotechnol 117(1):1–17

Pan X, Saddler JN (2013) Effect of replacing polyol by organosolv and kraft lignin on the property and structure of rigid polyurethane foam. Biotechnol Biofuels 6(1):12

Pan X, Arato C, Gilkes N (2005) Biorefining of softwoods using ethanol organosolv pulping: preliminary evaluation of process streams for manufacture of fuel-grade ethanol and coproducts. Biotechnol Bioeng 90(4):473–481

Pan X, Gilkes N, Kadla J (2006a) Bioconversion of hybrid poplar to ethanol and co-products using an organosolv fractionation process: optimization of process yields. Biotechnol Bioeng 94 (5):851–861

Pan X, Kadla JF, Ehara K, Gilkes N, Saddler JN (2006b) Organosolv ethanol lignin from hybrid poplar as a radical scavenger: relationship between lignin structure, extraction conditions, and antioxidant activity. J Agric Food Chem 54(16):5806–5813

Panagiotopoulos IA, Chandra RP, Saddler JN (2012) A two-stage pretreatment approach to maximise sugar yield and enhance reactive lignin recovery from poplar wood chips. Bioresour Technol 130:570–577

Papa G, Varanasi P, Sun L (2012) Exploring the effect of different plant lignin content and composition on ionic liquid pretreatment efficiency and enzymatic saccharification of Eucalyptus globulus L.Mutants. Bioresour Technol 117:352–359

Park N, Kim HY, Koo BW, Yeo H, Choi IG (2010) Organosolv pretreatment with various catalysts for enhancing enzymatic hydrolysis of pitch pine (Pinus rigida). Bioresour Technol 101(18):7046–7053

Perez-Pimienta JA, Lopez-Ortega MG, Varanasi P (2013) Comparison of the impact of ionic liquid pretreatment on recalcitrance of agave bagasse and switchgrass. Bioresour Technol 127:18–24

Playne MJ (1984) Increased digestibility of bagasse by pretreatment with alkalis and steam explosion. Biotechnol Bioeng 26:426–433

Poornejad N, Karimi K, Behzad T (2013) Improvement of saccharification and ethanol production from rice straw by NMMO and [BMIM][OAc] pretreatments. Ind Crops Prod 41:408–413

Prado R, Erdocia X, Serrano L, Labidi J (2012) Lignin purification with green solvents. Cellul Chem Technol 46(3–4):221–225

Pu YQ, Jiang N, Ragauskas AJ (2007) Ionic liquids as a green solvent for lignin. J Wood Chem Technol 27:23–33

Pye EK, Lora JH (1991) The AlcellTM process: a proven alternative to kraft pulping. Tappi J 74 (3):113

Qiu Z, Aita GM, Walker MS (2012) Effect of ionic liquid pretreatment on the chemical composition, structure and enzymatic hydrolysis of energy cane bagasse. Bioresour Technol 117:251–256

Reshamwala S, Shawky BT, Dale BE (1995) Ethanol production from enzymatic hydrolysates of AFEX-treated coastal Bermuda grass and switchgrass. Appl Biochem Biotechnol 51(52):43–55

Roberts BA, Strauss CR (2005) Toward rapid, green, predictable microwave-assisted synthesis. Acc Chem Res 38:653–661

Rollin JA, Zhu Z, Sathisuksanoh N, Zhang Y-HP (2011) Increasing cellulose accessibility is more important than removing lignin: a comparison of cellulose solvent-based lignocellulose fractionation and soaking in aqueous ammonia. Biotechnol Bioeng 108:22–30

Ruan D, Zhang L, Zhou J, Jin H, Chen H (2004) Structure and properties of novel fibers spun from cellulose in NaOH/thiourea aqueous solution. Macromol Biosci 4:1105–1112

Ryu DDY, Lee SB (1982) Effect of compression milling on cellulose structure and on enzymatic hydrolysis kinetics. Biotechnol Bioeng 24:1047–1067

Saddler JN, Ramos LP, Breuil C (1993) Steam pretreatment of lignocellulosic residues (Chap. 3). In: Saddler JN (ed) Bioconversion of forest and agricultural plant residues. CAB International, Oxford, UK, pp 73–92

Saha BC, Biswas A, Cotta MA (2008) Microwave pretreatment enzymatic saccharification and fermentation of wheat straw to ethanol. J Biobased Mater Bioenergy 2:210–217

Sarkar N, Ghosh SK, Bannerjee S, Aikat K (2012) Bioethanol production from agricultural wastes: an overview. Renew Energy 37(1):19–27

Sathitsuksanoh N, Zhu Z, Templeton N, Rollin J, Harvey S, Zhang Y-HP (2009) Saccharification of a potential bioenergy crop, Phragmites australis (common reed), by lignocellulose fractionation followed by enzymatic hydrolysis at decreased cellulase loadings. Ind Eng Chem Res 48:6441–6447

Sathitsuksanoh N, Zhu Z, Ho T-J, Bai M-D, Zhang Y-HP (2010) Bamboo saccharification through cellulose solvent-based biomass pretreatment followed by enzymatic hydrolysis at ultra-low cellulase loadings. Bioresour Technol 101:4926–4929

Sathitsuksanoh N, Zhu ZG, Wi S, Zhang Y-HP (2011) Cellulose solvent based biomass pretreatment breaks highly ordered hydrogen bonds in cellulose fibers of switchgrass. Biotechnol Bioeng 108:521–529

Sathitsuksanoh N, Zhu Z, Zhang Y-HP (2012a) Cellulose solvent and organic solvent-based lignocellulose fractionation enabled efficient sugar release from a variety of lignocellulosic feedstocks. Bioresource Technol 117:228–233

Sathitsuksanoh N, George A, Zhang Y-HP (2012b) New lignocellulose pretreatments using cellulose solvents: a review. J Chem Technol Biotechnol 88:169–180

Shafiei M, Karimi K, Taherzadeh MJ (2010) Pretreatment of spruce and oak by N-methylmorpholine-N-oxide (NMMO) for efficient conversion of their cellulose to ethanol. Bioresour Technol 101:4914–4918

Shafiei M, Karimi K, Taherzadeh MJ (2011) Technoeconomical study of ethanol and biogas from spruce wood by NMMO-pretreatment and rapid fermentation and digestion. Bioresour Technol 102(17):7879–7886

Shafizadeh F, Bradbury AGW (1979) Thermal degradation of cellulose in air and nitrogen at low temperatures. J Appl Polym Sci 23:1431–1442

Shafizadeh F, Lai YZ (1975) Thermal degradation of 2-deoxy-Darabino-hexonic acid and 3-deoxy-D-ribo-hexono-1,4-lactone. Carbohydr Res 42:39–53

Sheehan J, Aden A, Paustian K, Killian K, Brenner J, Walsh M, Nelson R (2003) Energy and environmental aspects of using corn stover for fuel ethanol. J Ind Ecol 7:117–146

Shi J, Chinn MS, Sharma-Shivappa RR (2008) Microbial pretreatment of cotton stalks by solid state cultivation of Phanerochaete chrysosporium. Bioresour Technol 99:6556–6564

Shin SJ, Sung YJ (2008) Improving enzymatic hydrolysis of industrial hemp (Cannabis sativa L.) by electron beam irradiation. Radiat Phys Chem 77:1034-1038

Shuai L, Yang Q, Zhu JY (2010) Comparative study of SPORL and dilute-acid pretreatments of spruce for cellulosic ethanol production. Bioresour Technol 101(9):3106–3114

Singh P, Suman A, Tiwari P (2008) Biological pretreatment of sugarcane trash for its conversion to fermentable sugars. World J Microbiol Biotechnol 24:667–673

Sivers MV, Zacchi G (1995) A techno-economical comparison of three processes for the production of ethanol from pine. Bioresour Technol 51:43–52

Socha AM, Parthasarathi R, Shi J, Pattathil S, Whyte D, Bergeron M, George A, Tran K, Stavila V, Venkatachalam S, Hahn MG, Simmons BA, Singh S (2014) Efficient biomass pretreatment using ionic liquids derived from lignin and hemicellulose. PNAS. doi:10.1073/pnas.1405685111

Striegel AM (1997) Theory and applications of DMAc/LiCl in the analysis of polysacharrides. Carbohydr Polym 34:267–274

Su MY, Tzeng WS, Shyu YT (2010) An analysis of feasibility of bioethanol production from Taiwan sorghum liquor waste. Bioresour Technol 101(17):6669–6675

Sun Y, Cheng J (2002) Hydrolysis of lignocellulosic materials for ethanol production: a review. Bioresour Technol 83:1–11

Swatloski RP, Spear SK, Holbrey JD (2002) Dissolution of cellulose with ionic liquids. J Am Chem Soc 124:4974–4975

Szijártó N, Kádár Z, Varga E, Thomsen AB, Costa-Ferreira M, Réczey K (2009) Pretreatment of reed by wet oxidation and subsequent utilization of the pretreated fibers for ethanol production. Appl Biochem Biotechnol 155(1–3):386–396

Takacs E, Wojnarovits L, Foldavary C, Hargagittai P, Borsa J, Sajo I (2000) Effect of combined gamma-irradiation and alkali treatment on cotton–cellulose. Radiat Phys Chem 57:339

Tengerdy RP, Nagy JG (1988) Increasing the feed value of forestry waste by ammonia freeze explosion treatment. Biol Wastes 25:149–153

Torget RW, Werdene P, Grohmann K (1990) Dilute acid pretreatment of two short-rotation herbaceous crops. Appl Biochem Biotechnol 24(25):115–126

Torget RW, Himmel M, Grohmann K (1992) Dilute acid pretreatment of two short-rotation herbaceous crops. Appl Biochem Biotechnol 34(35):115–123

Tucker MP, Kim KH, Newman MM, Nguyen QA (2003) Effects of temperature and moisture on dilute-acid steam explosion pretreatment of corn stover and cellulase enzyme digestibility. Appl Biochem Biotechnol 10:105–108

Ungurean M, Fitigau F, Paul C, Ursoiu A, Peter F (2011) Ionic liquid pretreatment and enzymatic hydrolysis of wood biomass. World Acad Sci Eng Technol 76:387–391

Uppal SK, Kaur R, Sharma P (2011) Optimization of chemical pretreatment and acid saccharification for conversion of sugarcane bagasse to ethanol. Sugar Tech 13(3):214–219

Vancov T, Alston AS, Brown T, McIntosh S (2012) Use of ionic liquids in converting lignocellulosic material to biofuels. Renew Energy 45:1–6

Varga E, Schmidt AS, Réczey K, Thomsen AB (2003) Pretreatment of corn stover using wet oxidation to enhance enzymatic digestibility. Appl Biochem Biotechnol 104(1):37–50

Ventura SPM, Santos LDF, Saraiva JA, Coutinho JAP (2012) Concentration effect of hydrophilic ionic liquids on the enzymatic activity of Candida antarctica lipase B. World J Microbiol Biotechnol 28:2303–2310

Verma P, Watanabe T, Honda Y, Watanabe T (2011) Microwave assisted pretreatment of woody biomass with ammonium molybdate activated by H_2O_2. Bioresour Technol 102:3941–3945

Vidal PF, Molinier J (1988) Ozonolysis of ligninsImprovement of in vitro digestibility of poplar sawdust. Biomass 16:1–17

Villaverde JJ, Ligero P, de Vega A (2010) Miscanthus x giganteus as a source of biobased products through organosolv fractionation: a mini review. Open Agric J 4:102–110

Vlasenko EY, Ding H, Labavitch JM, Shoemaker SP (1997) Enzymatic hydrolysis of pretreated rice straw. Bioresour Technol 59:109–119

Wang Y, Zhao Y, Deng Y (2008) Effect of enzymatic treatment on cotton fiber dissolution inNaOH/ureasolution at cold temperature. Carbohydr Polym 72:178–184

Wang Q, Wu Y, Zhu S (2011a) Use of ionic liquids for improvement of cellulosic ethanol production. BioResources 6(1):1–2

Wang K, Yang HY, Xu F, Sun RC (2011b) Structural comparison and enhanced enzymatic hydrolysis of the cellulosic preparation from Populus tomentosa Carr. by different cellulose-soluble solvent systems. Bioresour Technol 102:4524–4529
Wang C, Zhou F, Yang Z (2012) Hydrolysis of cellulose into reducing sugar via hot-compressed ethanol/water mixture. Biomass Bioenergy 42:143–150
Wasserscheid P, Keim W (2000) Ionic liquids—new solutions for transition metal catalyst. Angew Chem Int Ed 39:3773–3789
Weerachanchai P, Leong SSJL, Chang MW, Ching CB, Lee JM (2012) Improvement of biomass properties by pretreatment with ionic liquids for bioconversion process. Bioresour Technol 111:453–459
Weil JR, Sariyaka A, Rau SL, Goetz J, Ladisch CM, Brewer M (1997) Pretreatment of yellow poplar wood sawdust by pressure cooking in water. Appl Biochem Biotechnol 68:21–40
Weil J, Brewer M, Hendrickson R, Sariyaka A, Ladisch MR (1998) Continuous pH monitoringduring pretreatment of yellow poplar wood sawdust by pressure cooking in water. Appl Biochem Biotechnol 70–72:99–111
Wilson NS, Sarko CR, Roth GP (2004) Development and applications of a practical continuous flow microwave cell. Org Proc Res Dev 8(3):535–538
Wojciak A, Pekarovicova A (2001) Enhancement of softwood kraft pulp accessibility for enzymatic hydrolysis by means of ultrasonic irradiation. Cellul Chem Technol 35:361–369
Wright JD (1998) Ethanol from biomass by enzymatic hydrolysis. Chem Eng Prog 84(8):62–74
Wu J, Zhang J, He J (2004) Homogeneous acetylation of cellulose in a new ionic liquid. Biomacromolecules 5:266–268
Wu H, Mora-Pale M, Miao J, Doherty TV, Linhardt RJ, Dordick JS (2011) Facile pretreatment of lignocellulosic biomass at high loadings in room temperature ionic liquids. Biotechnol Bioeng 108(12):2865–2875
Wyman CE, Dale BE, Elander RT, Holtzapple M, Ladisch MR, Lee YY (2005a) Comparative sugar recovery data from laboratory scale application of leading pretreatment technologies to corn stover. Biores Technol 96(18):2026–2032
Wyman CE, Dale BE, Elander RT, Holtzapple MT, Ladisch MR, Lee YY (2005b) Coordinated development of leading biomass pretreatment technologies. Bioresour Technol 96:1959–1966
Wyman CE, Dale BE, Elander RT (2009) Comparative sugar recovery and fermentation data following pretreatment of poplar wood by leading technologies. Biotech Prog 25:333–339
Xie RQ, Li XY, Zhang YF (2012) Cellulose pretreatment with 1-methyl-3-methylimidazoliumdimethylphosphate for enzymatic hydrolysis. Cellul Chem Technol 46(5–6):349–356
Yang Q, Pan X (2012) Pretreatment of Agave americana stalk for enzymatic saccharification. Bioresour Technol 126:336–340
Yang B, Wyman CE (2004) Effect of xylan and lignin removal by batch and flow through pretreatment on enzymatic digestibility of corn stover cellulose. Biotechnol Bioeng 86:88–95
Yang B, Wyman CE (2008) Pretreatment: the key to unlocking low-cost cellulosic ethanol. Biofuels, Bioprod Biorefin 2(1):26–40
Yang CP, Shen ZQ, Yu GC (2008) Effect and after effect of gamma radiation pretreatment on enzymatic hydrolysis of wheat straw. Bioresour Technol 99:6240–6245
Yang F, Hanna M, Sun R (2012) Value-added uses for crude glycerol—a byproduct of biodiesel production. Biotechnol Biofuels 5, article 13
Yao RS, Hu HJ, Deng SS, Wang H, Zhu HX (2011) Structure and saccharification of rice straw pretreated with sulfur trioxide micro-thermal explosion collaborative dilutes alkali. Bioresour Technol 102(10):6340–6343
Youssef BM, Aziz NH (1999) Influence of gamma-irradiation on the bioconversion of rice straw by Trichoderma viride into single cell protein. Cytobios 97:171–183
Yu J, Zhang J, He J, Liu Z, Yu Z (2009) Combinations of mild physical or chemical pretreatment with biological pretreatment for enzymatic hydrolysis of rice hull. Bioresour Technol 100:903–908

Yuan D, Rao K, Varanasi S, Relue R (2012) Available method and configuration for fermenting biomass sugars to ethanol using native Saccharomyces cerevisiae. Bioresour Technol 117:92–98

Zavrel M, Bross D, Funke M, Buchs J, Spiess AC (2009) High-throughput screening for ionic liquids dissolving (ligno-) cellulose. Bioresour Technol 100:2580–2587

Zeitsch KJ (2000) The chemistry and technology of furfural and its many by-products, sugar series, vol 13. Elsevier, New York

Zhang Y-HP (2008) Reviving the carbohydrate economy via multi-product biorefineries. J Ind Microbiol Biotechnol 35:367–375

Zhang Y-HP, Lynd LR (2006) A functionally based model for hydrolysis of cellulose by fungal cellulase. Biotechnol Bioeng 94:888–898

Zhang H, Wu J, Zhang J (2005) 1-Allyl-3-methylimidazolium chloride room temperature ionic liquid: a new and powerful nonderivatizing solvent for cellulose. Macromolecules 38:8272–8277

Zhang Y-HP, Cui J, Lynd LR, Kuang LR (2006) A transition from cellulose swelling to cellulose dissolution by o-phosphoric acid: evidence from enzymatic hydrolysis and supramolecular structure. Biomacromolecules 7:644–648

Zhang XY, Xu CY, Wang HXJ (2007a) Pretreatment of bamboo residues with Coriolus versicolor for enzymatic hydrolysis. J Biosci Bioeng 104:149–151

Zhang XY, Yu HB, Huang HY (2007b) Evaluation of biological pretreatment with white rot fungi for the enzymatic hydrolysis of bamboo culms. Int Biodeterior Biodegrad 60:159–164

Zhang Y-HP, Ding S-Y, Mielenz JR, Elander R, Laser M, Himmel M, McMillan JD, Lynd LR (2007c) Fractionating recalcitrant lignocelluloses at modest reaction conditions. Biotechnol Bioeng 97:214–223

Zhang S, Marechal F, Gassner M, Perin-Levasseur Z, Qi W, Ren Z, Yan Y, Farvat D (2009) process modelling and integration of fuel ethanol production from lignocellulosic biomass based on double acid hydrolysis. Energy Fuels. doi:10.1021/ef801027x

Zhang Z, O'Hara IM, Doherty WOS (2012) Pretreatment of sugarcane bagasse by acid-catalysed process in aqueous ionic liquid solutions. Bioresour Technol 120:149–156

Zhang T, Kumar R, Wyman CE (2013a) Sugar yields from dilute oxalic acid pretreatment of maple wood compared to those with other dilute acids and hot water. Carbohydr Polym 92:334–344

Zhang DS, Yang Q, Zhu JY, Pan XJ (2013b) Sulfite (SPORL) pretreatment of switchgrass for enzymatic saccharification. Bioresour Technol 129:127–134

Zhao X, Liu D (2012) Fractionating pretreatment of sugarcane bagasse by aqueous formic acid with direct recycle of spent liquor to increase cellulose digestibility-the Formiline process. Bioresour Technol 117:25–32

Zhao H, Kwak JH, Wang Y, Franz JA, White JM, Holladay JE (2007) Interactions between cellulose and N-methylmorpholine-N-oxide. Carbohydr Polym 67(1):97–103

Zhao X, Cheng K, Liu D (2009) Organosolv pretreatment of lignocellulosic biomass for enzymatic hydrolysis. Appl Microbiol Biotechnol 82(5):815–827

Zheng YZ, Lin HM, Tsao GT (1995) Supercritical carbon-dioxide explosion as a pretreatment for cellulose hydrolysis. Biotechnol Lett 17:845–850

Zheng YZ, Lin HM, Tsao GT (1998) Pretreatment for cellulose hydrolysis bycarbon dioxide explosion. Biotechnol Prog 14:890–896

Zhi S, Yu X, Wang X, Lu X (2012) Enzymatic hydrolysis of cellulose after pretreated by ionic liquids: focus on one-pot process. Energy Procedia 14:1741–1747

Zhu SD (2008) Perspective used of ionic liquids for the efficient utilization of lignocellulosic materials. J Chem Technol Biotechnol 83:777–779

Zhu JY, Pan XJ (2010) Woody biomass pretreatment for cellulosic ethanol production: technology and energy consumption evaluation. Bioresour Technol 101(13):4992–5002

Zhu SD, Yu ZN, Wu YX (2005a) Enhancing enzymatic hydrolysis of rice straw by microwave pretreatment. Chem Eng Commun 192:1559–1566

Zhu YM, Lee YY, Elander RT (2005b) Optimization of dilute-acid pretreatment of corn stover using high-solids percolation reactor. Appl Biochem Biotechnol 121:325–327

Zhu SD, Wu YX, Chen QM, Yu N, Wang CW, Jin SW (2006) Dissolution of cellulose with ionic liquids and its application: a mini-review. Green Chem 8:325–327

Zhu Z, Sathitsuksanoh N, Vinzant T, Schell DJ, McMillan JD, Zhang Y-HP (2009) Comparative study of corn stover pretreated by dilute acid and cellulose solvent-based lignocellulose fractionation: enzymatic hydrolysis, supramolecular structure, and substrate accessibility. Biotechnol Bioeng 103:715–724

Zhu JY, Pan X, Ronald S, Zalesny J (2010a) Pretreatment of woody biomass for biofuel production: energy efficiency, technologies, and recalcitrance. Appl Microbiol Biotechnol 87:847–857

Zhu JY, Zhu W, Obryan P (2010b) Ethanol production from SPORL-pretreated lodgepole pine: preliminary evaluation of mass balance and process energy efficiency. Appl Microbiol Biotechnol 86(5):1355–1365

Zhua JY, Wang GS, Pan XJ, Gleisner R (2009) Specific surface to evaluate the efficiencies of milling and pretreatment of wood for enzymatic saccharification. Chem Eng Sci 64:474–485

Zwart RWR, Boerrigter H, Van der Drift A (2006) The impact of biomass pretreatment on the feasibility of overseas biomass conversion to fischer-tropsch products. Energy Fuels 20 (5):2192–2197

Chapter 5
Summary of Biomass Pretreatment Methods

Abstract Advantages, limitations, and disadvantages of various pretreatment processes for lignocellulosic biomass are presented in this chapter. Combination of two or more pretreatment processes is proven to be efficient when compared with single pretreatment process alone in terms of reducing sugar yield and lignin removal from different biomasses.

Keywords Pretreatment · Lignocellulosic biomass · Advantages · Limitations · Disadvantages

Several types of materials are found to be suitable for the production of biofuels. It must be stressed that it is not always possible to transfer the results of pretreatment from one type of biomass material to another. Furthermore, one technology that is effective for a particular type of biomass material might not be suitable for another material. All of the pretreatment methods discussed can lead to a high yield of glucose from cellulose as long as suitable feedstock and sufficient enzyme activities are used in hydrolysis. It is not the enzymatic accessibility that actually matters in the overall cost of biomass processing. However, the other factors such as enzyme dosing, total recovery of sugars (especially hemicellulose sugars), equipment, and energy cost, and so forth, can vary dramatically among the different types of pretreatment technologies and will result in different overall process economics. Also, it is obvious that the solid substrates obtained from different pretreatment methods vary greatly in composition and properties, which shows that the optimal enzyme recipes could be very different for each of the substrates. An in-depth understanding of the substrates and how they affect the enzyme functions is very important.

Various pretreatment processes for lignocellulosic biomass, and their advantages and disadvantages, are summarized in Table 5.1. The choice of the pretreatment technology used for a particular biomass is dependent on the composition of biomass and the byproducts produced as a result of pretreatment (Elander et al. 2009; McMillan 1994; Johnson and Elander 2007; Bensah and Mensah 2009; Hsu 1996;

P. Bajpai, *Pretreatment of Lignocellulosic Biomass for Biofuel Production*,
SpringerBriefs in Green Chemistry for Sustainability,
DOI 10.1007/978-981-10-0687-6_5

Table 5.1 Summary of various processes for the pretreatment of lignocellulosic feedstocks

Pretreatment process	Advantages	Limitations and disadvantages
Acid hydrolysis	Hydrolyzes hemicellulose to xylose and other sugars; alters lignin structure	High cost; equipment corrosion; formation of toxic substances
	Increase in porosity/increased enzymatic hydrolysis	Generation of furfural/hydroxymethyl furfural/need for recycling/costly
Alkaline hydrolysis	Removes hemicelluloses and lignin; increases accessible surface area	Long residence times required; irrecoverable salts formed and incorporated into biomass
		Formation of salts of calcium and magnesium
Organosolv	Hydrolyzes lignin and hemicelluloses; pure lignin obtained and used as value added product	Solvents need to be drained from the reactor, evaporated, condensed, and recycled; high cost; solvents inhibit enzymatic hydrolysis
AFEX	Increases accessible surface area, removes lignin and hemicellulose to an extent; does not produce inhibitors for downstream processes; decrystallization of cellulose	Not efficient for biomass with high lignin content; costly
Ammonia treatment	Removal of lignin/decrystallizing cellulose	Removal of ammonia/costly
Mechanical comminution	Reduces cellulose crystallinity	Power consumption usually higher than inherent biomass energy
Steam explosion	Causes hemicellulose degradation and lignin transformation; cost-effective	Destruction of a portion of the xylan fraction; incomplete disruption of the lignin-carbohydrate matrix; generation of compounds inhibitory to microorganisms
CO_2 explosion	Increases accessible surface area; cost-effective; does not cause formation of inhibitory compounds	Does not modify lignin or hemicelluloses
Pyrolysis	Produces gas and liquid products	High temperature; ash production
Ozonolysis	Reduces lignin content; does not produce toxic residues	Large amount of ozone required; expensive
	Lignin is damaged; cellulose/hemicellulose unaltered	
Biological	Degrades lignin and hemicelluloses; low energy requirements	Rate of hydrolysis is very low; a part of fermentable sugars are utilized as carbon source
Wet oxidation	Treatment of wastes	Costly
Microwave Treatment	Cheap/generates less pollution	Degradation of cellulose/hemicellulose

Based on Kumar et al. (2009), Chaturvedi and Verma (2013)

Wooley et al. 1999; Eggeman and Elander 2005). These factors substantially affect the costs associated with a pretreatment method.

Several researchers have made a comparison of various pretreatment methods for lignocellulosic feedstocks (Wyman et al. 2005a, b; Wyman 2007; Rosgaard et al. 2007; Silverstein et al. 2007). Rosgaard et al. (2007) evaluated the effectiveness of different types of pretreatment procedures, i.e., acid or water impregnation followed by steam explosion versus extraction with hot water, on wheat straw and barley. The pretreatments were compared after enzyme treatment using cellulase enzyme. The acid or water impregnation followed by steam explosion of barley straw was found to be the best pretreatment in terms of the resulting glucose concentration in the liquid hydrolysate after enzymatic hydrolysis.

Silverstein et al. (2007) examined the effectiveness of sulfuric acid, sodium hydroxide, hydrogen peroxide, and ozone pretreatments for the conversion of cotton stalks to ethanol. Solids from sulfuric acid, sodium hydroxide, and hydrogen peroxide pretreatments showed significant lignin degradation and/or high sugar availability and therefore were hydrolyzed rapidly by cellulsae enzymes—Celluclast 1.5 L and Novozym 188 from Novozymes. Pretreatment with sulfuric acid resulted in the highest xylan reduction (95.2 % for 2 % acid, 121 °C/15 psi, 90 min) but the lowest cellulose-to-glucose conversion during hydrolysis (23.9 %). Sodium hydroxide pretreatment resulted in the highest level of delignification (65.6 % for 2 % NaOH, 121 °C/15 psi, 90 min) and cellulose conversion (60.8 %). Pretreatment with hydrogen peroxide resulted in significantly reduced delignification (maximum of 29.5 % for 2 %, 121 °C/15 psi, 30 min) and cellulose conversion (49.8 %) in comparison to sodium hydroxide pretreatment. Ozone did not cause any significant changes in xylan, glucan, or lignin contents over time.

Wyman et al. (2005b) studied various pretreatment technologies for corn and reported that different methods yield different results. Therefore, the choice of pretreatment technology for a particular material depends on which components of the biomass required to be altered.

None of the processes discussed are found to be efficient enough to give 100 % yield of reducing sugars from different types of biomasses. The efficiency of the process depends mainly on the type of biomass used as raw material, its structure, and the lignin content (Mckendry 2002; Cao et al. 2012). Combination of two or more pretreatment processes is found to be efficient when compared with single pretreatment process alone in terms of reducing sugar yield and lignin removal from different biomasses. Therefore, combinations of pretreatment processes are being studied by several researchers for increasing the yield of reducing sugars (Cara et al. 2006; Yu et al. 2009; Miura et al. 2012).

An increased use of biofuels would contribute to sustainable development by reducing emissions of greenhouse gas and the use of nonrenewable resources. Lignocellulosic feedstocks, including forestry and agricultural residues instead of traditional feedstocks such as starch crops, could prove to be an ideally inexpensive and amply available source of sugar for fermentation into transportation fuels. Cellulose crystallinity, protection by lignin, accessible surface area, and sheathing by hemicellulose contribute to the resistance of cellulose in biomass to hydrolysis.

The biomass pretreatment and the intrinsic structure of the biomass itself are basically responsible for its subsequent hydrolysis. The conditions used in the selected pretreatment method will affect various characteristics of the substrate. This in turn, governs the susceptibility of the substrate to hydrolysis and the subsequent fermentation of the liberated sugars. Therefore, pretreatment of biomass is a very important step in the synthesis of biofuels from lignocellulosic feedstocks. There is a critical requirement to understand the fundamentals of various processes, which can help in making a suitable choice depending on the biomass structure and the hydrolysis agent.

References

Bensah EC, Mensah M (2009) Chemical pretreatment methods for the production of cellulosic ethanol: technologies and innovations Hindawi Publishing Corporation. Int J Chem Engg 2013:21, Article ID719607. http://dx.doi.org/10.1155/2013/719607

Cao W, Sun C, Liu R, Yin R, Wu X (2012) Comparison of the effects of five pretreatment methods on enhancing the enzymatic digestibility and ethanol production from sweet sorghum bagasse. Bioresour Technol 111:215–221

Cara C, Ruiz E, Ballesteros I, Negro MJ, Castro E (2006) Enhanced enzymatic hydrolysis of olive tree wood by steam explosion and alkaline peroxide delignification. Proc Biochem 41(2):423–429

Chaturvedi V, Verma P (2013) An overview of key pretreatment processes employed for bioconversion of lignocellulosic biomass into biofuels and value added products. 3. Biotech 3:415–431. doi:10.1007/s13205-013-0167-8

Eggeman T, Elander R (2005) Process and economic analysis of pretreatment technologies. Bioresour Technol 96(18):2019–2025

Elander RT, Dale BE, Holtzapple M, Ladisch MR, Lee YY, Mitchinson C, Saddler JN, Wyman CE (2009) Summary of findings from the biomass refining consortium for applied fundamentals and innovation (CAFI): corn stover pretreatment. Cellulose 16:649–659

Hsu TA (1996) Pretreatment of biomass. In: Wyman CE (ed) Handbook on bioethanol, production and utilization. Taylor & Francis, Washington, pp 179–212

Johnson DK, Elander RT (2007) Pretreatments for enhanced digestibility of feedstocks. In: Himmel ME (ed) Biomass recalcitrance. Blackwell, London, pp 436–453

Kumar P, Barrett DM, Delwiche MJ, Stroeve P (2009) Methods for pretreatment of lignocellulosic biomass for efficient hydrolysis and biofuel production. Ind Eng Chem Res 48:3713–3729

Mckendry P (2002) Energy production from biomass (part 1) overview of biomass. Bioresour Technol 83:37–46

McMillan JD (1994) Pretreatment of lignocellulosic biomass. In: Himmel ME, Baker JO, Overend RP (eds) Enzymatic conversion of biomass for fuels production. American Chemical Society, Washington, pp 292–324

Miura T, Lee SH, Inoue S, Endo T (2012) Combined pretreatment using ozonolysis and wet-disk milling to improve enzymatic saccharification of Japanese cedar. Bioresour Technol 126:182–186

Rosgaard L, Pedersen S, Meyer AS (2007) Comparison of different pretreatment strategies for enzymatic hydrolysis of wheat and barley straw. Appl Biochem Biotechnol 143:284–296

Silverstein RA, Chen Y, Sharma-Shivappa RR, Boyette MD, Osborne J (2007) A comparison of chemical pretreatment methods for improving saccharification of cotton stalks. Bioresour Technol 98:3000–3011

Wooley R, Ruth M, Glassner D, Sheehan J (1999) Process design and costing of bioethanol technology: a tool for determining the status and direction of research and development. Biotechnol Prog 15:794–803

Wyman CE (2007) What is (and is not) vital to advancing cellulosic ethanol. Trends Biotechnol 25 (4):153–157

Wyman C, Dale B, Elander R, Holtzapple M, Ladisch M, Lee Y (2005a) Coordinated development of leading biomass pretreatment technologies. Bioresour Technol 96(18): 1959–1966

Wyman CE, Dale BE, Elander RT, Holtzapple M, Ladisch MR, Lee YY (2005b) Comparative sugar recovery data from laboratory scale application of leading pretreatment technologies to corn stover. Bioresour Technol 96:2026–2032

Yu J, Zhang J, He J, Liu Z, Yu Z (2009) Combinations of mild physical or chemical pretreatment with biological pretreatment for enzymatic hydrolysis of rice hull. Bioresour Technol 100: 903–908

Chapter 6
Future Perspectives

Abstract Some major challenges in the area of lignocellulosic biomass pretreatment are presented. Future research required is also discussed.

Keywords Lignocellulosic biomass · Pretreatment · Future research · Challenges

The effect of greenhouse gasses on the climate change has been recognized as a serious environmental threat. Serious efforts are being made for the search of sustainable more efficient and environmental friendly technology to prevent such emission. Production of ethanol from lignocellulosics has received much attention since the last decade due to enormous potential for conversion of renewable biomaterials into biofuels. A major impediment in this technology is the presence of lignin, which inhibits hydrolysis of cellulose and hemicellulose. This has resulted in extensive research in the development of various pretreatment processes for the treatment of lignocellulosic biomass. The pretreatment step plays a very important role in a lignocellulosic biorefinery process.

Various research groups and companies at various levels, usually with financial support from national governments and public bodies (example Swedish Energy Agency, Danish Ministry of Energy, US Department of Energy/Agriculture, and Canadian Sustainable Development Technology Canada) and multinational institutions such as the European Union have developed several proprietary cellulosic ethanol production configuration and technologies. Chemical pretreatment of lignocellulosic biomass due to its high reactivity at mild conditions forms the basics of these technologies.

Table 6.1 gives profiles of some of the main projects undertaken or under construction/development underpinned by breakthrough pretreatment, hydrolysis, and fermentation technologies, as well as process integration and optimization. The pretreatment of feedstocks to improve biodegradability to simple sugars has been the subject of intensive research worldwide with a focus on maximizing sugar yields at high solid loads and at the lowest economic and environmental costs. Widely known and emerging chemical pretreatment methods have been reviewed with regard to process description, advantages, drawbacks, and recent innovations

P. Bajpai, *Pretreatment of Lignocellulosic Biomass for Biofuel Production*,
SpringerBriefs in Green Chemistry for Sustainability,
DOI 10.1007/978-981-10-0687-6_6

Table 6.1 Selected large-scale cellulosic ethanol plants based on chemical pretreatment

Plant / Pretreatment; feedstock
SEKAB Örnsköldsvik, Sweden Two stage dilute H_2SO_4/SO_2; pine chips
Abengoa bioenergy, Salamanca, Spain; York, NE, USA; Kansas, USA Acid impregnation + steam explosion; wheat and barley straw, corn stover, wheat straw, switchgrass
BioGasol Ballerup, Denmark Dilute acid/steam explosion or wet explosion; wheat straw and bran, corn stover, garden wastes, energy crops and green wastes
Procethol 2G, Futurol Pomacle, France Wheat straw, switchgrass, green waste, miscanthus, vinasses
Izumi Biorefinery Japan Arkenol; cedar, pine, and hemlock
INEOS Florida, USA Thermochemical; municipal solid waste and so forth
ZeaChem Oregon, USA Chemical; hybrid poplar, corn stover, and cob
Logos Technologies California, USA Colloid milling; corn stover, switchgrass, and wood chips
BlueFire Mississippi, USA Concentrated acid (Arkenol); wood waste, municipal solid waste
Weyland AS, Norway Concentrated acid; corn stover, sawdust, paper pulp, switchgrass
Borregaard Norway Acidic/neutral sulphite; wheat straw, eucalyptus, spruce
Queensland University of Technology, Queensland, Australia Acid, alkaline, steam explosion, ionic liquid; sugarcane bagasse
Praj Industries, India Thermochemical; corn cob, sugarcane bagasse
Lignol Energy Corporation, Canada Organosolv; wood, agricultural residues
Blue sugars, Wyoming, USA Acid, thermomechanical; pine
Petrobras/Blue sugars; Brazil Acid, thermomechanical; sugarcane bagasse
Dupont, Danisco Cellulosic Ethanol (DDCE), Iowa, USA NH_3 steam recycled; corn stover
COFCO/SINOPEC/Novozyme, Zhaodong, China Steam explosion (with acid impregnation); corn stover

Based on Bensah and Mensah (2013)

employed to offset inherent challenges. Cellulosic ethanol is close to commercialization but there are still technical, environmental, and economic challenges associated with biomass pretreatment, hydrolysis, and fermentation. No solvent has been found to work best for all biomass and such optimized methods and process conditions for various materials need to be examined and developed further. Some major challenges of the chemical pretreatment include the following:

- Requirement of extensive size reduction
- Handling biomass at high solids concentration
- Corrosion
- Solvent costs, and recovery
- Environmental pollution from solvents, by-products, and waste from reactions.

Nevertheless, the challenges mentioned above are being tackled via several interventions, particularly, the application of novel solvents and the combination of different chemical methods with physicochemical and biological pretreatments to obtain higher sugar yields, reduced use of costly solvents, lower enzyme dose, milder process conditions, recovery, and use of biomass components in pristine forms, and also improvements in environmental sustainability. Presently, several efforts are being made to develop new technologies to further reduce the cost of pretreatment and generate less toxic chemicals, higher sugar yield, and higher-value by-products. The choice of the pretreatment technology depends on several factors, such as the type of biomass, the value of by-products, and the process complexity (Chaturvedi and Verma 2013). The combination of different methods may yield more positive effects in the future. Extensive research has been done on the development of advanced pretreatment technologies to produce more digestible biomass in order to ease bioconversion of biomass into cellulosic ethanol. An ideal cost-effective pretreatment method should have the following characteristics (Hsu et al. 1996; Yang and Wyman 2008; Drapcho et al. 2008):

(1) Maximum fermentable carbohydrate recovery
(2) Minimum inhibitors generated as a result of carbohydrate degradation during pretreatment
(3) Reduced environmental impact
(4) Lower demand of post-pretreatment processes such as washing, neutralization, and detoxification
(5) Reduced use of water and chemicals
(6) Reduced capital cost for reactor
(7) Moderately reduced energy input
(8) Relatively high treatment rate
(9) Production of high value-added by-products.

Therefore, the future research on pretreatment should be focused on the following areas (Zheng et al. 2009):

(1) Reduction of water and chemical use
(2) Recovery of carbohydrates and value-added by-products to improve the economic feasibility
(3) Development of clean delignification yielding benefits of co-fermentation of hexose and pentose sugars with improved economics of pretreatment

(4) Basic understanding of pretreatment mechanisms and also the relationship between the biomass structure features and enzymatic hydrolysis
(5) Reduction of the formation of inhibitors such as furfural, 5-hydroxymethyl furfural and acetic acid which could significantly inhibit enzymatic hydrolysis and fermentation of biomass (Hsu et al. 1996; Yang and Wyman 2008).

Like most established industrial processes, a possible major step towards bioprocessing of lignocellulosics on an industrial scale is creating or finding markets for by-products of biomass pretreatment technologies (Agbor et al. 2011). Multiple or combinatorial pretreatments have the ability to enhance biomass digestibility and operate under various conditions to maximize selective product recovery, while reducing the generation of inhibitory carbohydrate degradation products. As pretreatment is the second most expensive unit cost in the conversion of lignocellulosic biomass to ethanol (NREL 2002), it calls for systematic analysis of pretreatment process dynamics and their by-products as means to reduce cost in designing a cost competitive process (Agbor et al. 2011). Lignol Innovations Burnaby, British Columbia, Canada has demonstrated its unique and economical integrated process technology for biorefining abundant and renewable lignocellulosic biomass feedstocks into fuel ethanol, pure lignin and other valuable co-products. Lignol's unique delignification pretreatment Organosolv process fractionates or separates woody biomass into cellulose, hemi-cellulose and lignin. The cellulose and hemi-cellulose are enzymatically hydrolysed into sugars. These sugars are fermented into ethanol which can be distilled and dehydrated to fuel-grade ethanol. The high purity lignin can be processed into a great variety of high value products. This innovation is a key solution for producing ethanol and high value products from low-value feedstocks, while providing an alternative to reliance on fossil fuels.

In the United States, at DOE bioenergy research centres, AFEX pretreated spent grains and lignocellulosics are being considered as feed for livestock. Although industrial production of biofuels has preceded a detail understanding of pretreatment, optimization of integrated biorefining processes requires strong coordinated research that will, delineate pretreatment chemistries and their effects on feedstocks and also fermentation yields of fuels and co-products.

References

Agbor VB, Cicek N, Sparling R, Berlin A, Levin DB (2011) Biomass pretreatment: fundamentals toward application. Biotechnol Adv 29:675–685

Bensah EC, Mensah M (2013) Chemical pretreatment methods for the production of cellulosic ethanol: technologies and innovations. Int J Chem Eng 2013:21 (Article ID719607)

Chaturvedi V, Verma P (2013) An overview of key pretreatment processes employed for bioconversion of lignocellulosic biomass into biofuels and value added products. Biotech 3:415–431

Drapcho CM, Nhuan NP, Walker TH (2008) Ethanol production. In: Biofuels Engineering Process Technology. McGraw-Hill, New York, pp 158–174

Hsu TA (1996) Pretreatment of biomass. In: Wyman CE (ed) Handbook on Bioethanol, production and Utilization. Taylor & Francis, Washington, pp 179–212

NREL (2002) Enzyme sugar-ethanol platform project. National Renewable Energy Laboratory, Golden, Co. http://www1.eere.energy.gov/biomass/pdfs/stage2_overview.pdf

Yang B, Wyman CE (2008) Pretreatment: the key to unlocking low-cost cellulosic ethanol. Biofuels Bioprod Bioref 2:26–40

Zheng Y, Pan Z, Zhang R (2009) Overview of biomass pretreatment for cellulosic ethanol production. Int J Agric Biol Eng 2(3):51–68

Index

P. Bajpai, *Pretreatment of Lignocellulosic Biomass for Biofuel Production*,
SpringerBriefs in Green Chemistry for Sustainability,
DOI 10.1007/978-981-10-0687-6

Printed in the United States
By Bookmasters